Origin of Inertia

Extended Mach's Principle and Cosmological Consequences

Amitabha Ghosh

Apeiron
Montreal

Published by Apeiron
4405, rue St-Dominique
Montreal, Quebec H2W 2B2 Canada
http://redshift.vif.com

First Published 2000

Canadian Cataloguing in Publication Data

Ghosh, Amitabha, 1941 Dec. 3-
Origin of inertia : extended Mach's principle and cosmological consequences

Includes Index
ISBN 0-9683689-3-X

1. Inertia (Mechanics) 2. Mach's principle. 3. Celestial mechanics. I. Title.

QC137.G48 2000 521 C00-900906-X

Cover design by Priyadarshi Patnaik

Dedicated
to

the late Dr. Toivo Jaakkola

Contents

Foreword

ONE OF the enduring mysteries of classical and modern physics arises from the fact that the value obtained for the gravitational-mass ratio of two particles compared in a weighing experiment is identical to the value obtained for the inertial-mass ratio of the same two particles compared in a collision experiment (to within experimental error). For this reason we speak of the equivalence between inertial and gravitational mass, and tend to use the concepts interchangeably. However, whilst practitioners of science and engineering have been content to accept this equivalence as a matter of fact, there has been little rational understanding of it. Amitabha Ghosh, the author of this book, provides us with a particular unifying perspective for this baryonic dichotomy—and, in the process, unifies many other disparate phenomena as well.

Amitabha Ghosh is a very fine mechanical engineer, as can be judged from his positions as a Senior Fellow of the Alexander von Humboldt Foundation, in Germany, from 1977 to 1995, as a Fellow of the Institution of Engineers (India) and of the three major academies in India—the Indian National Academy of Engineering, New Delhi, the Indian National Science Academy, New Delhi and the Indian Academy of Science, Bangalore; furthermore, he is fluent in German, English, Bengali and Hindi in addition to (he says!) having a working knowledge of the very old language, Sanskrit. Any conversation with Amitabha will, sooner or later, come round to a discussion (actually, more of a one-way 'dialogue' for a second party like myself!) of this language, and of how it has all the appearance of being a language that was designed from its basics upwards—rather than, as is virtually every other language, being an evolved language.

I would imagine that it is Amitabha's attention to the intellectual detail—as exemplified by his interest in Sanskrit—that has informed his life's work as an outstanding mechanical engineer and more recently, his directorship of IIT Kharagpur, and the contents of this book.

But first, consider the IITs (or Indian Institutes of Technology): until my friendship with Amitabha, I had (to my insular shame) never heard of the IITs—but a measure of their amazing quality can be had by considering the following statistics: Every year, about 140,000 of the best school students in the whole of India (population 1 billion) apply for about *two thousand* places in six IITs (I prefer not to make any comparison with what happens in the UK). IIT Kharagpur (est. 1951) was the first and the largest of the six IITs, and so, by the primary measure of student quality, Amitabha is the director of one of the world's leading higher-educational institutions.

And so to the book! I would begin by saying that it stands as a timely reminder to all of us that fundamental thought in 'mathematical physics' does not have, as a pre-requisite, necessary knowledge of group theory, quantum mechanics, general relativity, string theory, quantum algebras, knot theory, etc etc. What fundamental thought *does* require is a thorough familiarity with the phenomena, combined with a proper appreciation of the ways in which the received views *fail* to address these phenomena, further combined with a willingness to suppose that, perhaps, those who have gone before, no matter how revered, have managed to overlook some crucial point. This book provides a vivid illustration of how such brave and sideways thinking can work in the general area of space, time, inertia and gravitation.

The material developed here conforms closely to Tom Phipp's philosophy of the 'covering theory'—the idea that the most stable and reliable way for physical theory to progress is to move from a successful (but limited) theory to a theory which 'covers' (or includes as a special case) the earlier limited theory. In essence, Amitabha has adopted the position that Newtonian Mechanics (that veritable *Queen* of sciences) is a good theory (after proper attention is paid to the idea of 'inertial frame') and that its child, Newtonian gravitation theory, is also good *at first-order*. That is, Amitabha shows that, in order to obtain a gravitation theory which works on the largest spatial and longest temporal scales, we emphatically do not need to scrap the whole Newtonian paradigm, replacing this whole with a completely new conceptual basis; on the contrary, all that is needed is a creative *tweak* which invokes the spirit of Mach and incorporates a little thinking of a kind which many engineers might see as pure common sense.

The result is a simple and elegant theory which at once provides a rational basis for understanding Newton's second law (for those who don't see it as a definition) and, simultaneously, gives Mach a genuine and belated 'engineering role' in the finer gravitational phenomena.

In detail, the book begins with an excellent overview of Newtonian ideas, and follows this with a brief but accessible discussion on, usually overlooked, fundamental difficulties associated with these ideas. This is followed by a discussion of Mach's Principle and of Dennis Sciama's 1950s ideas of acceleration-induced 'inertial induction' as a means of realizing the Principle. Finally, in this developmental phase, the shortcomings of the Sciama approach are described and a means of circumventing these is proposed—the notion of a velocity-induced inertial induction mechanism in addition to the Sciama mechanism. There then follows detailed analyses of many types of gravitationally interactive systems.

Apart from the lucid simplicity of the book's arguments, what impresses most is the ability of such a simple no free-parameter model to explain a very wide variety of phenomena—some of which are currently anomolous and not explained by any other theory (e.g., the secular acceleration of Phobos), some of which have ad-hoc explanations in the standard theories (e.g., the flat rotation curves of spiral galaxies and the invocation of 'dark matter' as the cure-all) and some of which have a ready explanation in standard theory (e.g., the cosmic redshift seen as evidence of the big-bang theory of universal origin).

In short, this work does what any *good* theory should—it provides a simple unifying mechanism to explain a wide variety of phenomena and makes strongly testable predictions—I have in mind the secular retardation of Mars (actually predicted in the text) and the new generation of gyro experiments currently planned as the definitive test of that much under-tested standard theory, general relativity—which, of course, by the lights of the ungenerous, has already failed the galactic dynamics test!

As a final serious point, what a much finer world it would be if books like this could find their way easily into undergraduate physics and mathematical physics courses as the definitive inoculation against the disease of hubris which is the curse of contemporary University theoretical physics around the world.

Read and contemplate!

David Roscoe
Applied Mathematics Department
Sheffield University
Sheffield UK

Preface

The preface of a scientific work is usually considered unimportant by most readers. Only a few (not necessarily the serious readers) may take interest in this prelude, primarily to see how the author was motivated to write the book. Some may also want to check if their names figure in the acknowledgements. This is not a usual sort of scientific presentation. This is a monograph in which a mechanical engineer has ventured to extend and modify fundamental concepts of physics with a view to solving some of the longstanding unresolved problems of astrophysics and cosmology. This makes this preface not only very much in order, but perhaps, absolutely essential.

Let me explain how I made this adventurous digression from my own professional field of study. This unusual volume contains the results of about twelve years' (1983-1995) work and solitary effort on my part. It started with a desire to explain the origin of inertia to the students of my second year Dynamics class. Some anticipatory attempts to introduce certain modifications to the existing formalism led to startling results, which surprised my colleagues in the Physics Department of the Indian Institute of Technology, Kanpur. They found the results to be very intriguing, though difficult to reconcile with the framework of conventional physics. As soon as a few initial conclusions were corroborated through existing experimental and/or observational results, some reputable cosmologists and astronomers suggested that I apply the proposed modifications to more and more cases. I persisted, and found that the theory was not only able to yield quantitatively correct results in each case, but in many cases resolved certain unexplained phenomena. Then I realised for the first time that I could not set the whole matter aside but should try to develop a systematic theory on the basis of what I propose and what I have accomplished. However, I must mention that only rarely does one encounter clear phenomena in nature where only one mechanism is operative. Since the magnitudes involved in such phenomena are very small, it is difficult to come to a definite conclusion. It has been possible to identify a few cases where conclusions can be drawn with a reasonable degree of confidence, and the observations support the proposed theory.

The primary theme of this monograph is a theory in which Newton's static gravitational interaction has been replaced by a new dynamic model. According to this theory the interactive gravitational force between two objects depends not only on the separation but also on the relative velocity and acceleration between the interacting bodies. In a sense it is an extension of Mach's Principle, and could be termed the Extended Mach's Principle. According to Mach's Principle a force acts on an accelerating object due to its interaction with the matter present in the rest of the universe. In the Extended Mach's Principle such an interactive force acts on a body due to its velocity (in the mean rest frame of the universe) also. This force has been termed the Cosmic Drag. The initial Chapters offer some general discussions and highlight certain interesting features of the fundamental problem of motion. It has been demonstrated that in a relational framework an absolute character can be assigned to displacement, and hence it is meaningful to talk about an absolute frame of reference in an infinite, non-evolving and quasistatic universe satisfying the Perfect Cosmological Principle. A brief account of the basic difficulties in the Newtonian formulation of mechanics is then presented, followed by a historical account of how the earlier researchers tried to resolve these issues through various suggestions for modifications in Newton's laws. The next chapter presents Sciama's attempt to quantify Mach's Principle and his model of acceleration dependent inertial induction. Next, this is extended to include a velocity-dependent inertial induction term. Though very small, the effect of the velocity dependent inertial induction term introduces some fundamental changes in the basic framework of mechanics, leading to a modified law of motion. The exact equivalence of gravitational and inertial masses emerges as a natural consequence of the dynamic gravitational interaction.

Another startling result of this modification is the emergence of a cosmic drag term, whereby all objects are subjected to a drag force depending on the velocity with respect to the mean-rest-frame of the infinite, homogeneous and quasistatic universe. Though not easily open to detection by any experiment because of its extremely small magnitude, this cosmic drag term gives rise to the observed cosmological redshift without invoking any expansion hypothesis. This eliminates the need for introducing a Big Bang to start the universe. Here I anticipate a criticism that I am questioning a cosmological theory accepted by the majority of mainstream physicists around the world. The Big Bang Theory has become so popular now that very few remember it is still only a hypothesis. Scientists who believe that the search for alternatives to Big Bang cosmology should not be restricted are quite fewer in number. I am presenting this alternative model with the belief that the scope of scientific research is always wider than many people think. The proposed model yields very good results in a number of other phenomena of different types. This is very important, as these phenomena are unrelated, and the proposed model does not have any adjustable free parameters. The proposed theory and any model of the extended version of inertial induction can be further tested on new observations with higher accuracy. Since this model is likely to open new vistas, other researchers can take up the necessary work to verify the correctness of the proposed theory.

The most difficult part of the whole exercise is deciding on an appropriate title of this monograph. An attempt to make it technically precise can render it unfamiliar and uninteresting. On the other hand, a catchy and attractive title may be imprecise and go against the true scientific spirit. I have tried to strike a balance, but whether I have been successful in my endeavour can be judged by the readers. My primary objective is to attract the attention of prospective serious readers. However, it is the subject matter and the text, not the title, which ultimately decides whether the initial interest will be sustained or not as a reader goes through the book.

This monograph was first conceived in the departure lounge of Montreal international airport in summer 1993, and the main impetus came from C. Roy Keys. His dedication and contribution to the cause of science surpass those of many professional scientists. The idea of the monograph was renewed and further strengthened during discussions with Dr. David Roscoe of Sheffield University and Dr. A. K. Gupta of Allahabad University in January 1997 when we assembled at Bangalore to participate in an international workshop. However, it would not have been possible for me to complete the task without the enthusiastic and invaluable help from Dr. S. Banerjee of the Electrical Engineering Department of the Indian Institute of Technology, Kharagpur. A major part of the credit for this monograph being completed should go to his untiring effort and kind help. Thanks are also due to Dr. P. Pattanaik of the Humanities and Social Science Department of the Indian Institute of Technology, Kharagpur, for designing the cover page of this book. I received tremendous encouragement and help during the early stages of this work from Profs. H. S. Mani, Y. R. Waghmare, A. K. Mallik and R. Singh of the Indian Institute of Technology, Kanpur. Subsequently I had very useful and encouraging discussions with Prof. J. V. Narlikar, of Inter-University Centre for Astronomy and Astrophysics, Pune, Prof. A. K. Roy Chowdhury of Presidency College, Calcutta and C. V. Visveshwara of the Indian Institute of Astrophysics, Bangalore. Since 1990 I have been in contact with a number of astronomers and scientists from Europe, The UK, Canada, The USA and Brazil who showed keen interest in the model, and the experience and insight I gained through detailed discussions with them at various meetings and workshops during the period 1990 to 1995 helped me enormously to gain confidence in the matter. I am indebted to all of them. I also met the late Dr. Toivo Jaakkola for the first time in 1990 and a number of times after that. We had the opportunity to work together for a short while during Dr. Jaakkola's visit to the Indian Institute of Technology, Kanpur in April-May 1993. He represents the eternal tragedy of science where recognition comes long after one has left the scene without having the opportunity to see one's dream come true. I am also very thankful to Dr. Gayatri Sanini who spent about six months in perfecting many of my earlier analyses by checking many early estimates in order to arrive at more accurate results. Thanks are also due to Prof. T. N. Maulik, former Professor of Mathematics of Bengal Engineering College, Shibpur, India, who read the manuscript thoroughly and gave a number of valuable suggestions. The generous financial grant to prepare the manuscript provided by Prof. S. K. Sarangi, Dean, Continuing Education, from the Curriculum Development Cell of Indian Institute of Technology, Kharagpur, is gratefully acknowledged. I am truly indebted to my wife Mrs. Meena Ghosh who had to take all the trouble to run the household (no mean task in India for middle class families) so that I could have the necessary time to continue the work resulting in this monograph without any serious problem. At the end I would like to take this opportunity to thank many friends and students who took this unusual avocation of mine (outside the domain of my professional training and activities) with enough seriousness and provided me the necessary encouragement to continue.

Amitabha Ghosh

February 2000
Indian Institute of Technology
Kharagpur, West Bengal
India

Chapter 1

Introduction

1.1 Introduction

THE motion of objects was one of the first natural categories that attracted human attention, and the physical sciences started with the study of motion. Our environment is full of different sorts of motion, and an understanding of nature is not possible without a proper idea about the cause and effect relationship involving movement of objects. An understanding of the science of motion is very important from the point of view of philosophical interest; moreover, most of our present day science and technology would have been impossible without the development of the science of motion.

It is often believed that there is no scope for further development in this branch of science. This is far from the truth. Students come across the basic principles of kinematics, the three laws of motion and the law of universal gravitation at the high school level. They learn how to use the various rules and laws in solving problems in physics and mechanics, but most often without a deeper understanding of the science hidden behind these rules. At a later stage very few of them feel the need to go on to further studies of the fundamental aspects of the science of motion, as most engineering and scientific applications do not require such an effort. Yet it would be a mistake to think that an attempt to enquire into the deeper aspects of the science of motion is of no practical significance.

It is true that the laws of motion and universal gravitation lead to correct results for all engineering and scientific problems in the domain of non-relativistic classical mechanics. But many important issues may depend on those features of the laws of motion which can be ignored under normal circumstances. Therefore, a deeper examination of the laws of motion with a view to identifying the possible scope of modifications may prove quite a rewarding effort. With the progress of technology, detection of previously unknown effects may also become possible.

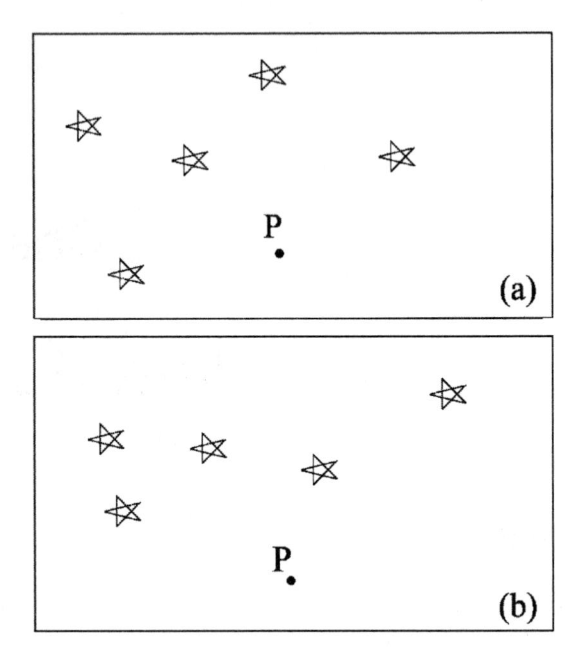

Figure 1.1: Particle P and objects in space at two different instants.

In the field of science of motion and mechanics one needs to be more careful, as in this field accurate measurements are still very difficult. Though the phenomenon of gravitation is most predominant in the formation of planetary, stellar and galactic systems, the accuracy with which we know the value of the gravitational constant, G, is the least compared to the other universal constants. This is mainly due to two reasons. The force of gravity, compared to the other medium and long-range forces, is extremely weak, and the measurements related to the science of motion are mostly of mechanical nature, for which the accuracy of measurement is relatively lower. Moreover, many features of the phenomenon related to motion and gravitation cannot be detected unless very large scale systems are considered. Terrestrial experiments are totally incapable of identifying such effects. For example, it is very difficult to say whether G is a constant or not. It is possible that it may decrease very slowly with distance, but the distance involved in any detectable change in G may be so large that it is practically impossible to demonstrate this by any direct experiment. However, such aspects of matter may come to light if the basic aspects of the laws of motion and gravitation are continually put under close scrutiny.

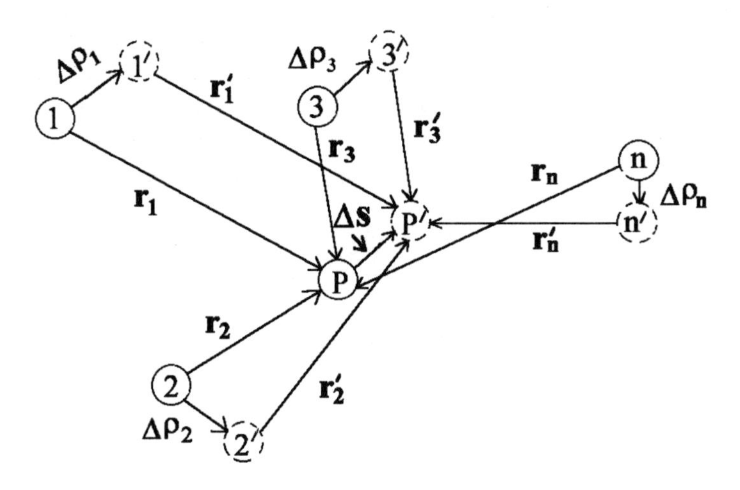

Figure 1.2: Relative displacements of objects.

1.2 Concept of Absolute Motion and Mean Rest Frame of the Universe

A major point which has to be resolved first is the concept of motion itself. Motion is loosely defined as change of position with time. As soon as the idea of position enters the picture the next important question is: "position with respect to what?" Newton believed that space is absolute and the concept of motion with respect to this absolute space is a viable postulate. However, most philosophers and scientists today feel that it is meaningless to talk about the position of a particle in a totally matter-free empty space. According to them the position of a body can be defined relative to other objects only. This results in a major difficulty, and quantification of motion becomes problematic. Figure 1.1a shows a point mass P and a few other objects in free space. If a snapshot of the same region is taken after some time, it may look somewhat like that shown in Fig.1.1b. Since all objects in this universe are in motion it is impossible to determine the amount of displacement of the particle P. In fact it is meaningless to talk about the displacement of the particle. This is considered to be the fundamental problem of motion,[1] and many philosophers and scientists believe that there is nothing like absolute motion and motion can be defined only in a relational manner.

However, the concept of absolute motion of a body is not meaningless if the universe is infinite, homogeneous and quasistatic. All objects in such a universe exhibit random motion of finite magnitude. It is shown below that using only the relative displacements of a particle P with respect to other objects (also moving) in the universe,

[1] Barbour, J.B., *Absolute or Relative Motion*, Cambridge University Press, 1989

a concept of displacement (of P) with respect to the mean rest frame of the universe is feasible. To begin with we can assume that every object can be considered to move with respect to an absolute space identified by the mean rest frame of the universe. Figure 1.2 shows a portion of the objects (infinite in number) in the universe and the particle P at two instants of time t and $(t + \Delta t)$. The position of P (in the assumed absolute space) shifts to P' during the interval Δt. All other objects also shift to new positions $1', 2', 3', \ldots\ldots n', \ldots\ldots$

The displacements of the objects and the point P are $\Delta\boldsymbol{\rho}_1, \Delta\boldsymbol{\rho}_2, \Delta\boldsymbol{\rho}_3, \ldots\Delta\boldsymbol{\rho}_n, \ldots$ and $\Delta\boldsymbol{s}$, respectively. Let the position of P be determined (measured) with respect to the objects in the universe at time t and $(t+\Delta t)$ as $\mathbf{r}_1, \mathbf{r}_2, \mathbf{r}_3, \ldots\mathbf{r}_n, \ldots$ and $\mathbf{r}'_1, \mathbf{r}'_2, \mathbf{r}'_3, \ldots$ $\mathbf{r}'_n, \ldots$, respectively. Then from Fig.1.2 we can write

$$\begin{aligned}
\mathbf{r}_1 + \Delta\mathbf{s} &= \Delta\boldsymbol{\rho}_1 + \mathbf{r}'_1 \\
\mathbf{r}_2 + \Delta\mathbf{s} &= \Delta\boldsymbol{\rho}_2 + \mathbf{r}'_2 \\
\mathbf{r}_3 + \Delta\mathbf{s} &= \Delta\boldsymbol{\rho}_3 + \mathbf{r}'_3 \\
&\ldots \\
\mathbf{r}_n + \Delta\mathbf{s} &= \Delta\boldsymbol{\rho}_n + \mathbf{r}'_n \\
&\ldots
\end{aligned}$$

Summing up both sides for N such reference objects when N is very large we get

$$\sum_{i=1}^{N} \mathbf{r}_i + N.\Delta\mathbf{s} = \Delta\rho + \sum_{i=1}^{N} \mathbf{r}'_i, \tag{1.1}$$

where $\Delta\boldsymbol{\rho}$ is the vector sum of all the random vectors $\Delta\boldsymbol{\rho}_i (i = 1, 2, \ldots N)$. It should be further noted that the magnitude of $\Delta\boldsymbol{\rho}$ will be given by the relation[2]

$$\Delta\rho = \sqrt{N}.\Delta\rho_{av}, \tag{1.2}$$

where $\Delta\rho_{av}$ is the average magnitude of $\Delta\rho_1, \Delta\rho_2, \Delta\rho_3, \ldots$ Using the above relation in (1.1) and rearranging the equation we get

$$\Delta\mathbf{s} = \frac{1}{N}\left(\sum_{i=1}^{N} \mathbf{r}'_i - \sum_{i=1}^{N} \mathbf{r}_i\right) \tag{1.3}$$

as $\Delta\rho/N \rightarrow 0$ when N tends to be very large. Equation (1.3) implies that when measuring the relative positions of P with respect to a large number of objects at the beginning and end of the time interval under consideration, the vector $\Delta\mathbf{s}$ can be considered to be the displacement of P with respect to the mean rest frame of the

[2]This is a standard random walk problem, and the derivation of the result can be found in most books on probability.

universe, and not with respect to any particular body. The idea of absolute motion[3] can be achieved by introducing the concept of the mean rest frame of the universe (provided it is infinite, homogeneous and quasistatic). After the existence of the cosmic microwave background radiation was detected the idea of "absolute motion" gained credibility. Observations show that this radiation is blueshifted (*i.e.*, the wavelength is reduced) in one direction of the sky and redshifted by an equal amount when observed in the opposite direction. This is due to the Doppler effect resulting from the motion of the Earth with respect to this isotropic radiation filling the universe. The magnitude of our velocity with respect to this radiation is about 260 km s^{-1}.

We will again return to the issue of the absolute and relational nature of the displacement of an object in the subsequent chapters. The main object of discussing this in the introductory chapter is to demonstrate the fundamental nature of the difficulty involved in defining these concepts, and to make the reader aware of this problem.

1.3 Laws of Motion and Universal Gravitation

1.3.1 The discovery of dynamics

One of the most interesting subjects of study is the history of the science of motion. It started with Aristotle and reached its crescendo with the publication of Newton's *Principia.* It may appear to be surprising to a student of science today that it took so much time to discover basic rules which appear to be so obvious now. Yet it must be remembered that it is not a simple matter to identify and investigate the basic features of motion, as most motions we observe are combinations of a number of motions of basic nature. The Earth's gravity, presence of friction, atmospheric effects, *etc.* introduce so many features in any motion that it becomes very difficult to determine the fundamental causes which govern motions of material objects.

The least complicated of all motions which could be observed by the philosophers and men of science in the past are the planetary motions and the motion of the Moon. Only when the true nature of these motions could be grasped by the human mind was the situation ripe for the discovery of the fundamental principles governing motions of objects. Therefore, it may not be incorrect to suggest that, had there been no dark sky at night (due to a second Sun in our system, for example) or no planets in our system, perhaps even to-day the laws of motion would have remained beyond the grasp of the human mind.

Johannes Kepler has to be given the credit for discovering the true nature of planetary motion. Once he was able to build the correct model of a heliocentric solar system and discover the laws obeyed by the planets in their motions around the Sun,

[3]However, it should be remembered that this absolute displacement does not require any concept of motion with respect to absolute space. The displacement is still obtained by taking into account the displacements of the particle with respect to the other objects present in the rest of the universe.

he could feel that the Sun must be responsible for the motions of the planets. Simultaneously Galileo had been making revolutionary discoveries, and identified acceleration as one of the main parameters in the science of motion. Therefore, as soon as he conceived the inertial properties of free motion it was not difficult for him to explain the motion of projectiles.[4] The true nature of the law of inertia was first proposed by Descartes. His proposition was almost identical to what Newton postulated in his Axiom I, *i.e.*, the popularly known first law of motion. The crucial point—that force produces acceleration—was first recognised by Huygens though the credit for the quantitative relationship goes to Newton. The concept of mass and the expression for the quantity of motion (momentum) was first obtained by Newton, and his second law was possible only after the development of these concepts. However, Huygens also developed the concept of momentum by analysing collision problems. It was Huygens who formulated the early ideas about conservation of linear momentum. The third law of motion is a discovery of Newton alone. Kepler realised that some kind of driving force emanates from the Sun, but his concept involved forces on the planets in the transverse direction. Only much later Hooke, Wren and Halley suggested that the force of the Sun on the planets was an attraction diminishing in intensity as the square of the distance. However, it required a genius like Newton to come up with a grand synthesis of all these ideas and provide a complete theory of universal gravitation. But this was possible primarily because Kepler had laid down the kinematic rules the planets' motions satisfy.

1.3.2 The laws of motion and universal gravitation

The currently popular text book forms of the three laws of motion are as follows:

First Law : Every body continues in its state of rest, or of uniform motion in a straight line, unless it is compelled to change that state by forces impressed upon it.

Second Law : The change of motion is proportional to the motive force impressed and is made in the direction of the straight line in which that force is impressed.

Third Law : To every action there is always opposed an equal reaction or, the mutual actions of two bodies upon each other are always equal, and in opposite directions.

The modern form in which the 2nd law is used in solving problems is well known and can be written as

$$\mathbf{F} = m\mathbf{a} \tag{1.4}$$

[4]Galileo did not conceive of rectilinear inertial motion as we know it today. He still believed that natural motions were circular in nature, and his law of inertia suggested that unhindered motions would be in circles. However, for the distances involved the effect of the circular motion was very small and his results were good.

where $\mathbf{F}$ is the impressed motive force, m is the inertial mass of the object and $\mathbf{a}$ is the resulting acceleration. However, the popular formula is valid only when m is constant.

Newton's law of universal gravitation is given by the attractive force between two mass particles. The expression for the law is as follows.

$$\mathbf{F}_{12} = \frac{G m_1 m_2}{r_{12}^3} \mathbf{r}_{12} \tag{1.5}$$

where $\mathbf{F}_{12}$ is the force on body 1 due to body 2, $\mathbf{r}_{12}$ is the position of body 2 with respect to body 1, m_1 and m_2 are the gravitational masses of bodies 1 and 2, respectively and G is a universal constant equal to $6.67 \times 10^{-11}\text{m}^3\text{kg}^{-1}\text{s}^{-2}$. In more popular form the force of mutual attraction between two point masses with gravitational masses m_1 and m_2 separated by a distance r is given by

$$F = \frac{G m_1 m_2}{r^2} \tag{1.6}$$

This mutual force of attraction acts along the line joining the two point masses.

1.3.3 The basic nature of motion

Though not important from the point of view of specific problems, the point which has created maximum confusion and contradiction in the minds of the philosophers and scientists is the basic nature of motion.[5] When an object moves, does it move with respect to other objects present or with respect to some entity like absolute space?[6] As long as *terra firma* had its immobile status, as in the Aristotelian school of thought, there was no problem. All motions were with respect to the Earth, just as we observe. The difficulty started once the firm ground was lost.

Before Newton, many scientists and philosophers, especially Descartes, believed in relational characteristics of motion. In their opinion, movements can be observed and felt only with respect to other objects. In the absence of any such background, motion can neither be felt nor have any meaning. It is difficult to say with any certainty whether Descartes emphasised this relational aspect in his famous book, *Principles of Philosophy*, to give a certain amount of legitimacy to Galileo's and others' theory that the Earth moves around the Sun without evoking too much hostility from the Church. However, this line of thinking was severely attacked by Newton. Since all objects possess a certain amount of motion it becomes very difficult to find a framework in which the law of inertia (and the second law of motion) is valid. If an object moves in a straight line with constant speed in the absence of any external force, with respect to

[5] The problem of the basic nature of motion has already been discussed in Sec.1.1 from a purely kinematic point of view. It is demonstrated how a relational description can be used to arrive at a result that can be considered to be valid with respect to the mean rest frame of the universe.

[6] For quantitative description a mean rest frame of the infinite quasistatic universe may be used to represent the "absolute space."

what does it describe a straight line? Newton also strongly criticised Descartes' idea of the philosophical nature of motion. He considered that 'motion' of a body has real existence, and for the reality of motion to be meaningful the existence of an absolute space becomes essential. The following passage from Newton's writings criticising Cartesian relativism reveals his thinking about the nature of motion:

> Lastly, that the absurdity of this position may be disclosed in full measure, I say that thence it follows that a moving body has no determinate velocity and no definite line in which it moves. And, what is worse, that the velocity of a body moving without resistance cannot be said to be uniform, nor the line said to be straight in which its motion is accomplished. On the contrary, there cannot be motion since there can be no motion without a certain velocity and determination.

Newton believed in the existence of absolute space, and according to his philosophy the inertial property of an object is its intrinsic property, independent of the presence of other material objects in the universe.

Subsequently Newton's concept of absolute space was attacked by Berkeley, Leibniz and Mach. According to them, motion has no meaning unless it is observed to exist with respect to other objects. They did not stop at that. They maintained that the resistance to acceleration of an object arises from its interaction with the matter present in the rest of the universe. Newton demonstrated his famous bucket experiment (to be discussed in detail in Chapter 3) to disprove the idea of the relational nature of motion, but later philosophers and scientists like Berkeley and Mach did not consider the experiment meaningful. According to them the effect of the relational movement with the material content of the bucket is too small to be observed, but it is present on the surface of the water. However, one cannot ignore the need for a framework for the laws of motion to be valid. The concept of inertial frame was developed to rescue the situation. These are the assumed frames of reference in which the laws of motion are valid. It will be shown in the next chapter that both groups had certain elements of truth in what they said. It will be also shown how it is possible for both the relational and absolute nature of motion to be valid simultaneously—a nice parallel to the wave-particle duality of matter.

Chapter 2

Difficulties with Newton's Laws of Motion

2.1 Introduction

NEWTON'S laws of motion and universal gravitation form the basic premise from which the study of physical sciences begins. Because of the phenomenal success of Newtonian mechanics, it was generally felt that these laws need not be subjected to any re-examination or critical analysis. Of course, now it is well known that Newtonian mechanics needs to be replaced by relativistic mechanics and quantum mechanics, depending on whether the speeds involved are comparable to that of light or the objects concerned are too small. But it is not considered necessary to examine the validity of Newton's laws in the domain of classical mechanics involving non-relativistic speeds. Since we learn to apply these laws in solving problems at very early stages—generally in the higher classes of the schools—the deeper and more subtle aspects go unnoticed by us. As already mentioned, even at the higher level the students learn the sophisticated analytical techniques only, but there is rarely any reopening of the basic issues, which remain hidden to most of them forever.

Hence, it is appropriate at this stage of our discussion to present these issues. This will not only help the reader to be familiar with the difficulties associated with the well-known and extensively used Newton's laws, but also be very useful for an understanding of the subsequent argument of this monograph. It should be pointed out at the very outset that the laws of motion and the universal law of gravitation are by no means completely free of difficulties, nor do they make correct predictions about or explain all phenomena governed by mechanics. One might be tempted to presume that incorporating relativistic effects may settle the difficulties, but that is not always the case. Moreover, in all cases the discrepancies between predictions and the corresponding observations are small and the basic framework of these laws is, naturally, taken to be correct. Yet sometimes very small modifications can have quite profound effects on some important aspects. For example, if from the very beginning Newton's second law had included a drag term (depending on the velocity of an object) of even an extremely small magnitude, whose effects it might be impossible to detect by any laboratory experiment, it could have accounted for the cosmological redshift without invoking the expansion hypothesis, and the accepted model of the universe would have been totally different. The problems associated with Newton's laws can be divided into two main groups. In the first group we can consider the difficulties associated with the basic philosophy of the law and the mystery and paradoxes arising out of the formulation, whereas the second group contains the difficulties associated with the predictions obtained from the application of these laws. In what follows these two groups of difficulties are presented.

2.2 Difficulties Associated with the Laws

The main problems in this group are of three types—ambiguity, mystery and paradox. Each of these is discussed below.

(a) Ambiguity : A considerable amount of ambiguity exists in the basic framework itself. Newton's laws are claimed to be valid only when the observations are made in an inertial frame of reference. Unfortunately there has never been any proper definition of an inertial frame of reference. Since all the cosmic objects are in perpetual motion, the search for an inertial frame has remained elusive and without success. Apart from this, the observation that motion has significance only when it is relative to another object, has cast a serious doubt on the existence of an absolute space.[1] Newton tried to resolve the issue by suggesting that inertial frames are those in which his proposed laws are valid. Thus the whole reasoning becomes circular in nature, and such a definition cannot be considered to be scientific. In fact it has been shown[2] that the logical outcome of the laws of motion proposed by Newton can be re-framed as "there exist inertial frames." This is a very serious ambiguity present in the framework of Newtonian mechanics. It is essential to arrive at a scientific way of defining the frame of reference in which the laws are valid.

(b) Mystery : A great mystery, unnoticed by many, surrounds the laws proposed by Newton. According to many eminent scientists this is one of the greatest mysteries in physical science. The property of a body, which governs its response

[1] However, the reader may recall the earlier discussion about how a mean rest frame of the universe can be conceived to make the idea of a preferential absolute frame of reference meaningful.

[2] Gyan Mohan, "Frames of Reference" in: *Lectures Delivered on the occasion of the Tercentenary of Newton's Principia*, IIT Kanpur, India. Feb.27-28, 1987 [Unpublished monograph].

to an externally applied force, is the *inertial mass*. On the other hand, the property which takes part in the gravitational interaction is the *gravitational mass*. In the framework of Newtonian mechanics there is no obvious link between the two phenomena, but it has been found that these two masses of any object are always equivalent. With suitable choice of the unit of mass these two can be made identically equal, *i.e.*,

$$m_i \equiv m_g \tag{2.1}$$

where m_i and m_g are the inertial and gravitational masses of a body. In a more generalised description of this equivalence one can state that the gravitational and inertial masses of an object are proportional to one another. Or,

$$m_i = \lambda m_g \tag{2.2}$$

where λ is the constant of proportionality. All the results would still remain valid if we put $G = \lambda^2 \times 6.67 \times 10^{-11}\text{m}^3\text{kg}^{-1}\text{s}^{-2}$. Thus the exact equivalence only implies that we have chosen $\lambda = 1$ and $G = 6.67 \times 10^{-11}\text{m}^3\text{kg}^{-1}\text{s}^{-2}$.

Einstein proposed to resolve the issue by developing a theory of gravitation, commonly known as the General Theory of Relativity. He postulated that the effects of acceleration and gravitational pull are indistinguishable from local observations. However, according to the model of dynamic gravitational interaction based on an Extended Mach's Principle, it can be shown that inertia is nothing but the manifestation of dynamic gravitation, and thereby the equivalence of inertial and gravitational masses can be better explained.

(c) Paradox : Another classical difficulty plagues Newton's inverse square law of universal gravitation. If the universe is assumed to be homogeneous, infinite (in space and time) and Euclidean,[3] then the potential U at a point for a particle of gravitational mass m_g is given by the following relation :

$$U = -\int_0^\infty \frac{Gm_g}{r} 4\pi r^2 \rho dr \tag{2.3}$$

where ρ is the average density of matter in the universe. When evaluated, the potential U tends to infinity if G is treated as a constant, as proposed.

The existence of the above mentioned gravitational paradox can be demonstrated in another way. Let us assume that the universe is homogeneous, infinite and Euclidean. A particle of gravitational mass m_g is at a point P as shown in Fig.2.1. Assume any point C and imagine a sphere with C as the centre and the length of the line $CP(= R)$ as its radius. The particle at P will be attracted by the matter

[3]"Euclidean universe" means one in which Euclidean geometry is valid.

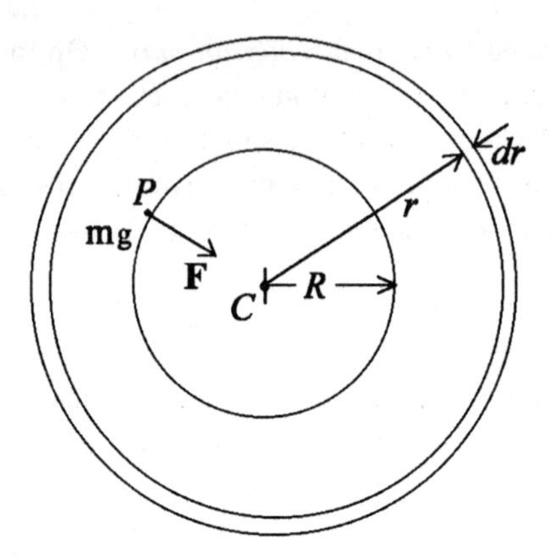

Figure 2.1: The gravitational paradox.

contained in this sphere according to the inverse square law, and the magnitude of this attracting force **F** will be given by

$$\begin{aligned} F &= \frac{4}{3}\pi R^3 \frac{G m_g}{\rho R^2} \\ &= \frac{4}{3}\pi G \rho R m_g \end{aligned} \tag{2.4}$$

This force **F** is directed towards the point C. The rest of the universe can be conceived to be constituted of concentric uniform spherical shells with C as their centre. One such shell of thickness dr and radius r is shown in Fig.2.1. According to the inverse square law, the force acting on a particle inside a uniform spherical shell is exactly equal to zero. So, the resultant force on the particle P due to the rest of the universe outside the sphere of radius R is zero. The net effect is that the particle is attracted to the point C by a force **F**. But the choice of C is totally arbitrary and, therefore, so is the case with the direction and magnitude of **F**.

It has been argued that the universe is neither infinite nor Euclidean according to the standard model of cosmology. Unfortunately, however, no observational evidence has been obtained so far to conclude that the universe is finite. At the same time the space-time of the universe has not been found to possess any curvature. As will be shown later, the problem can be resolved if G is found to diminish as the distance between the interacting particles increases.

2.3 Difficulties with the Predictions

The other group of difficulties is associated with the predictions from the laws of motion and universal gravitation proposed by Newton. Some of these difficulties are discussed below :

(a) Secular Motions : Newton's laws have failed to account for the total amount of observed secular motions of the celestial bodies. All the inner planets exhibit some excess perihelion rotation, of which that for the planet Mercury (about $43''$ per century) is most remarkable. Einstein's general theory of relativity is able to explain this excess advance, thought it can also be explained by assuming the Sun to possess a required degree of oblateness within the framework of Newtonian mechanics. It is now an established fact that the spin motion of the Earth is gradually slowing down, and the magnitude of this secular retardation is about 6×10^{-22} rad/s^2. The conventional explanation within the framework of Newtonian mechanics is the tidal friction due to the presence of the Moon. But this explanation encounters serious difficulties, because, according to this theory, the Moon should have been very close to the Earth 1000 million years ago, and that should have destroyed the natural satellite. More complicated and exotic mechanisms are being proposed to overcome the difficulty. Movement of the continents, post ice age elastic rebound of the Earth's crust (due to the reduction of polar ice caps), *etc.*, are examples of such theories. But it is still more difficult to explain the observed large secular acceleration of the Martian satellite, Phobos. Even the terrestrial artificial satellites show some features in their motions which cannot be accounted for by the conventional laws of motion. The most notable among these is the motion of the satellite LAGEOS which is tracked by laser beams with a very high degree of accuracy. In addition to these, this secular motion characteristics, the frequent occurrence of near commensurabilities in the orbital motions of the satellites of the major planets is also a mysterious problem.

(b) Transfer of Angular Momentum: It is now an accepted theory that the solar system evolved through the condensation and collapse of a nebular cloud. According to this theory the collapsing cloud rotates faster and faster as the angular momentum is conserved, and the planets finally form out of the matter ejected from the equatorial region of the spinning central body in the form of a protoplanetary disc. However, a major stumbling block in this hypothesis is that, even though the Sun contains 99.9% of the matter of the whole system, the angular momentum it possesses is only 0.5% of the total angular momentum of the solar system. This does not fit with the theory that the solar system evolved through the gravitational collapse of a rotating cloud, unless the central spinning body (*i.e.*, the Sun) lost its angular momentum to the planetary bodies. A number of mechanisms have been proposed to explain the transfer of solar angular momentum, all of which are valid

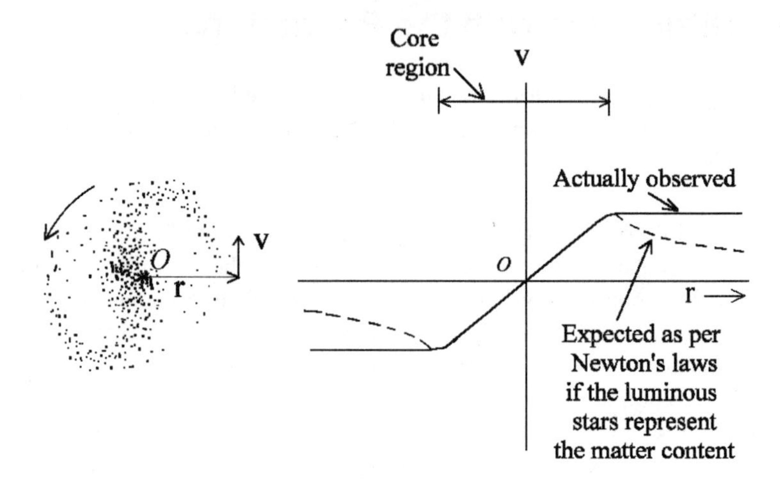

Figure 2.2: Typical velocity characteristics of stars in a spiral galaxy.

for the short pre-main sequence period. But the intensity required in each mechanism for the transfer of the observed amount of angular momentum is so high that their feasibility is questionable. Secondly, a similar odd distribution of angular momentum is observed in the planet-satellite systems also, and the proposed mechanisms cannot be active in such cases. Newton's laws do not provide any explanation, as the so-called tidal phenomenon is unable to transfer the required amount of angular momentum from the central spinning body.

(c) Galactic Rotation Curves and the Dark Matter Problem : More recently Newton's laws have faced a serious challenge. The orbital speeds of the stars in spiral galaxies are found not to obey Kepler's relations[4] and the law of universal gravitation. If the conventional Newtonian mechanics is valid on the galactic scale, then the orbital speeds of stars should gradually decrease with the increase in the distance of a star from the galactic centre (Fig.2.2). However, observation shows that beyond the central core region the orbital speeds do not exhibit the expected Keplerian fall-off, and remains almost constant for a very long distance (out to the extreme detectable edge) as indicated in Fig.2.2. The expected Keplerian fall-off is, of course, based on the assumption that luminosity of a region (or a star) is in accordance with the amount of matter present. The only way this can be explained within the frame-work of Newtonian mechanics is to assume that the actual matter content is far in excess of what is represented by the luminous bodies—*i.e.*, the stars. This is one of the considerations that have led to the theory of dark matter

[4]This observation is now well established and can be found in most of the modern books on astrophysics. Some detailed references are given in Chapter 8.

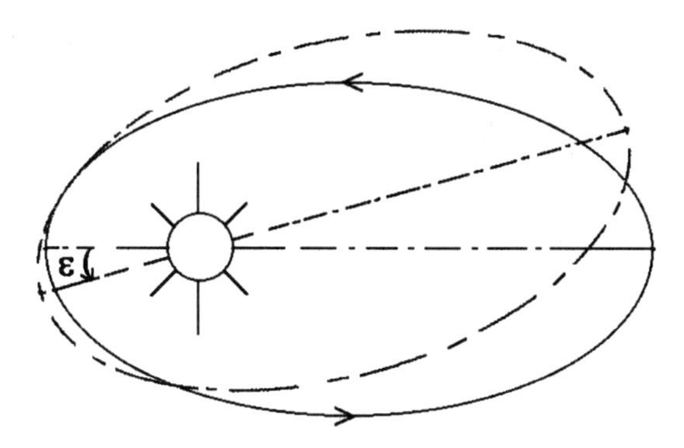

Figure 2.3: Advance of perihelion.

according to which about 90% of the matter present in a typical galaxy is dark and cannot be seen. Unfortunately, until now no definite idea is available about the nature of this dark matter. Moreover, an almost constant velocity characteristic is possible if and only if the matter in a galaxy is distributed in a unique manner. The flat rotation curve is a very common feature of almost all spiral galaxies and necessitates the existence of some servomechanism to distribute the matter in the required unique way. But Newtonian mechanics does not provide a suitable servomechanism for such a unique distribution of matter in spiral galaxies.

2.4 The Proposed Modifications of Newton's Laws

In view of these problems there have been many proposals for modifications of the laws of motion and universal gravitation presented by Newton. Some of these are intended primarily to explain the discrepancies between observations and theoretical predictions. Some researchers have tried to modify the basic philosophy about the nature of the phenomena in order to remove the problems associated with the laws themselves. In all cases the quantitative effects of the modifications are too small for verification by laboratory experiments. Besides, observational data on astronomical and astrophysical phenomena are seldom quite precise, *i.e.*, in most cases a large number of factors are simultaneously responsible for the discrepancies between the observed results and the theoretical predictions. This makes it very difficult to decide either in favour of or against a particular proposal to reshape a law on the anvil of observations. Of course, there are a few cases where the modifications suggested could not attract much support because their predictions agreed in a few situations but contradicted the observations in other situations.

Table 2.1: Excess perihelion rotation of planets

	Excess perihelion rotations in arc-seconds per century		
Orbiting Body	Modified Newtonian	Observed	According to general theory of relativity
Mercury	42.56	42.56	42.86
Icarus	9.95	9.8	10.02
Venus	8.54	8.4	8.6
Earth	3.8	4.6	3.83
Mars	1.34	1.5	1.35

It is interesting to note that Newton himself was the first to investigate the consequences of a modification to the law of universal gravitation.[5] In the *Principia*, Book I, under the heading "Proposition XLIV; Theorem XIV," Newton proves the following theorem:

> The difference of the forces, by which two bodies may be made to move equally, one in a fixed, the other in the same orbit revolving, varies inversely as the cube of their common altitudes.

In modern language this means that the perihelion of a planetary orbit advances (as shown in Fig.2.3) if the gravitational law is of the form

$$F = \frac{Gm_s m_p}{r^2} + \frac{C}{r^3} \tag{2.5}$$

Table 2.1 lists the estimated excess perihelion rotations, δ, of various planets[6] using the above equation (with a suitably chosen value of C), together with the corresponding astronomical values and predictions by Einstein's general theory of relativity. It is clear from the table that the modified Newtonian law can explain the excess perihelion shifts reasonably well.

The scientific community in the eighteenth and nineteenth centuries was overwhelmed by the power of Newtonian mechanics to explain the motions of the planets. The Newtonian program received a further boost from the discovery of Neptune. It is generally known that the existence and position of the planet was predicted with the help of Newtonian mechanics by analysing the data on the perturbations in the motion of Uranus. [7]

[5] Curé, J., *Galilean Electrodynamics,* Vol.2, 1991, p 43.

[6] Curé, J., *Galilean Electrodynamics*, Vol.2, 1991, p 43.

[7] However, it was later found that there was a considerable error in the calculation, and detection of the new planet was, in effect, a stroke of luck.

As the accuracy of observations improved, anomalies started appearing, the most noticeable being the excess perihelion motion of the planet Mercury. In the nineteenth century, researchers were eager to derive a modified form of Newton's law of gravitation which could explain the unexplained motions of the heavenly bodies. In 1846 Weber[8] proposed a theory of electrodynamics according to which the total repellent force between two charges e and e' of the same sign at a distance r is given by

$$F = \frac{ee'}{4\pi\epsilon r^2}\left[1 - \frac{1}{c^2}\left(\frac{dr}{dt}\right)^2 + \frac{2r}{c^2}\cdot\frac{d^2r}{dt^2}\right], \tag{2.6}$$

where c is the speed of light. Under his influence, contemporary astronomers started toying with the idea of a similar law of gravitation. Tisserand[9] proposed that the gravitational attraction between two particles of gravitational masses m_1 and m_2 at a distance r be also given by a law similar to (2.6) as follows :

$$F = \frac{Gm_1m_2}{r^2}\left[1 - \frac{1}{h^2}\left(\frac{dr}{dt}\right)^2 + \frac{2r}{h^2}\cdot\frac{d^2r}{dt^2}\right], \tag{2.7}$$

where h was considered to be the speed of propagation of gravitation. Taking h to be equal to the speed of light, he obtained the excess perihelion rotation of Mercury as $14''$ per century. Very recently Assis[10] has made further progress in this direction. He has expressed the equations of motion for the Sun and the planet as follows :

$$\mathbf{F}_s = m_s\mathbf{a}_s = \frac{Gm_sm_p}{r_{sp}^2}\hat{\mathbf{r}}_{sp}\left[1 + \frac{\xi}{c^2}\left(r_{sp}\ddot{r}_{sp} - \frac{\dot{r}_{sp}^2}{2}\right)\right] \tag{2.8}$$

and

$$\mathbf{F}_p = m_p\mathbf{a}_p = -\frac{Gm_sm_p}{r_{sp}^2}\hat{\mathbf{r}}_{sp}\left[1 + \frac{\xi}{c^2}\left(r_{sp}\ddot{r}_{sp} - \frac{\dot{r}_{sp}^2}{2}\right)\right], \tag{2.9}$$

where m_s and m_p are the masses of the Sun and planet, respectively, $\hat{\mathbf{r}}_{\mathbf{sp}}$ is a unit vector along the line from the Sun to the planet, r_{sp} is the distance and ξ is a constant. Using $M = m_s + m_p$, ρ as the radius vector from the Sun to planet and θ as the angular co-ordinate, the above two equations yield

$$\rho\ddot{\theta} + 2\dot{\rho}\dot{\theta} = 0 \tag{2.10}$$

and

$$\ddot{\rho} - \rho\dot{\theta}^2 = -GM\left[\frac{1}{\rho^2} + \frac{\xi}{c^2}\left(\frac{-\dot{\rho}^2}{2\rho^2} + \frac{\ddot{\rho}}{\rho}\right)\right]. \tag{2.11}$$

[8] Weber, W., *Leipzig Abhandl.*, 1846 : *Ann. d. Phys.*, lxxiii, 1848; English translation in Taylor's *Scientific Memoirs*, 1852, p.489.

[9] Tisserand, M.F., *Comptes Rendues de l'Académie des Sciences* (Paris), Vol.75, 1872, p.760

[10] Assis, A. K. T., *Found. Phys. Letters*. Vol.2, 1989, p. 301.

Defining $u = 1/\rho$ and $H = \rho^2\dot{\theta}$, (2.11) gives the following orbit equation:

$$\frac{d^2u}{d\theta^2} + u = GM\left[\frac{1}{H^2} - \frac{\xi}{c^2}\left\{\frac{1}{2}\left(\frac{du}{d\theta}\right)^2 + u\frac{d^2u}{d\theta^2}\right\}\right] \tag{2.12}$$

This equation can be solved by noting that the second and the third terms within the square bracket in the R.H.S. are much smaller than the first one. The solution can be written as

$$u(\theta) = u_0(\theta) + u_1(\theta), \tag{2.13}$$

with $|u_0| \gg |u_1|$ where $u_0(\theta)$ and $u_1(\theta)$ satisfy the following equations :

$$\frac{d^2u_o}{d\theta^2} + u_0 = \frac{GM}{H^2} \tag{2.14}$$

and

$$\frac{d^2u_1}{d\theta^2} + u_1 = -\frac{GM\xi}{c^2}\left[\frac{1}{2}\left(\frac{du_0}{d\theta}\right)^2 + u_0\frac{d^2u_0}{d\theta^2}\right]. \tag{2.15}$$

Solving (2.14), we obtain the classical result

$$u_0(\theta) = \frac{GM}{H^2} + A\cos(\theta - \theta_0), \tag{2.16}$$

where A and θ_0 are given by the initial conditions. Using (2.16) in (2.15) we get a particular solution for $u_1(\theta)$ as follows:

$$u_1(\theta) = a_1 + a_2(\theta - \theta_0)\sin(\theta - \theta_0) + a_3\cos^2(\theta - \theta_0), \tag{2.17}$$

where

$$a_1 = \frac{GMA^2}{2}\cdot\frac{\xi}{c^2} \tag{2.18}$$

$$a_2 = \frac{G^2M^2A}{2H^2}\cdot\frac{\xi}{c^2} \tag{2.19}$$

$$a_3 = -\frac{GMA^2}{2}\cdot\frac{\xi}{c^2}. \tag{2.20}$$

The peri- and aphelion positions are given by the condition $du/d\theta = 0$. The advance of the perihelion in one revolution is then given by

$$\frac{\epsilon}{100} = \Delta\theta = \pi\frac{\xi}{c^2}\frac{G^2M^2}{H^2} = \pi\frac{\xi}{c^2}\cdot\frac{GM}{a(1-e^2)} \tag{2.21}$$

where a is the semi-major axis and e is the eccentricity of the orbit. Taking $\xi = 6$, (2.21) becomes same as the value obtained by the general theory of relativity.

It is interesting to note that both Gauss and Riemann [11] also proposed modified versions of a gravitational law which were somewhat similar to the law given by (2.7). It was possible to produce some perihelion shifts. There were some *ad hoc* suggestions also which attempted to resolve the difficulty with modified versions of the law of gravitation. One such proposal was made by Bertrand[12] in 1873. He suggested the following gravitational law :

$$F = \frac{Gm_1m_2}{r^n} \tag{2.22}$$

where n is slightly greater than 2. Hall[12] showed that if n = 2.00000016, then the observed excess perihelion shift of Mercury could be explained. However, later it was shown[13] that this conflicts with the observations of the Moon's motion, and such hypotheses were not taken seriously.

More recently Maneff[14] has proposed that the mass of a body in the gravitational field of another body of mass M at a distance r is given by

$$m = m_0 \exp\left[\frac{GM}{c^2 r}\right], \tag{2.23}$$

here m_0 is, say, the proper mass. With this hypothesis the force of gravitational attraction can be expressed as

$$F = \frac{GMm_0}{r^2}\left[1 + \frac{3GM}{c^2 r}\right] \tag{2.24}$$

and the resulting excess perihelion rotation is the same as given by Einstein.

There were a few other proposals, also regarding the gravitational attraction, mostly during the last century, which are omitted here. However, with the rise of popularity of the general theory of relativity, the search for a modified gravitational force law to explain the excess perihelion shift has been more or less abandoned by mainstream researchers. As mentioned earlier, some recent works have attempted to investigate the possibility of the Sun's oblateness causing the observed excess perihelion shift.

The motivation to find a modified version of Newton's law of universal gravitation had another seed—to resolve the issue of gravitational paradox in an infinite, homogeneous Euclidean universe. It was Laplace[15] who was the first to suggest that gravitation was absorbed by matter situated between two interacting bodies of masses m_1 and m_2 separated by a distance r. He suggested the following modified form:

[11] Whittaker, E., *A History of the Theories of Aether and Electricity*, Vol. 1 and 2, Thomas Nelson & Sons Ltd., 1953.

[12] Hagiharia, Y., *Celestial Mechanics,* Part 1, MIT Press, 1972.

[13] Brown, E. W., *Monthly Notices of Royal Astronomical Society*, Vol.63, 1903, p.396.

[14] Maneff, C., *Zeitschrift für Physik*, Vol.34, 1930, p.766.

[15] Laplace, P. S., "Traité de Mécanique Céleste" in: *Oeuvres de Laplace*, Vol.5, Book 16, Chapter 4, Gauthier-Villars, Paris, 1880.

$$F = \frac{Gm_1m_2}{r^2}e^{-\lambda r}. \tag{2.25}$$

When the above relation is used the gravitational force $\mathbf{F_1}$ on a particle at the surface of a spherical volume uniformly filled with matter can be expressed (in non-dimensional form) as follows:

$$|F_1| = \int_0^1 x^2 \left[\int_{-1}^{1} \frac{e^{-\alpha\sqrt{1+x^2+2xy}}(1+xy)}{(1+x^2+2xy)^{3/2}} dy \right] dx. \tag{2.26}$$

The force acting on the same particle, $\mathbf{F_2}$, by all the infinite concentric uniform spherical shells outside the chosen spherical volume, can be expressed as

$$|F_2| = \int_1^\infty x^2 \left[\int_{-1}^{1} \frac{e^{-\alpha\sqrt{1+x^2-2xy}}(xy-1)}{(1+x^2-2xy)^{3/2}} dy \right] dx. \tag{2.27}$$

It can be shown by computation that $|F_1| = |F_2|$. Thus, $\mathbf{F_1} = -\mathbf{F_2}$ and an arbitrary particle is acted upon by no resultant force in an infinite, homogeneous universe. He also considered the effect of this modification on the motions of the heavenly bodies in the solar system. After analysing the results he placed an upper limit to λ as 7×10^{-18} m^{-1} so that the theory did not conflict with the observation. Seeliger[16] later took up this expression and showed that the gravitational paradox is resolved. In the sixties Pechlaner and Sexl[17] investigated a theory of gravity in which also the potential contains an additional exponentially decreasing term besides the Newtonian term. It will be shown later in this book that according to the model developed by the author the dynamic gravitational interaction decreases exponentially, and the value of λ is uniquely determined. Recently Kropotkin[18] has considered the Seeliger force model in connection with cosmology. The idea of absorption of gravity as proposed by Laplace (and followed by Seeliger) was again used by Bottlinger[19] to explain some discrepancies in the motion of the Moon noted by Newcomb in 1895. He proposed that these anomalies, which occurred mostly during the eclipses of the Moon, were due to the absorption of the Sun's gravity by the intervening Earth. He obtained a value of λ as 3×10^{-13} m^{-1} inside the Earth. Subsequently, more research has been conducted[20]

[16] Seeliger, H., *Astronomische Nachrichten*, Vol. 137, 1895, p. 129; Seeliger, H., *Über die Anwendung der Naturgesetze auf das Universum*, Berichte Bayer, Akad. Wiss., Vol.9, 1909.

[17] Pechlaner, E. and Sexl, H., *Communications in Math. Physics*. Vol.2, 1966, p.165.

[18] Kropotkin, P. N. — *Soviet Physics Doklade*, Vol.33 (2), 1988, p.85; Kropotkin, P. N. — *Soviet Physics Doklade*, Vol.34 (4), 1989, p.277; Kropotkin, P. N., *Apeiron*, Nos.9-10, 1991, p.91.

[19] Bottlinger, C. F. — *Astronomische Nachrichten*, Vol.1991, 1912, p. 147.

[20] Majorana, Q., *Comptes Rendues de l'Académie des Sciences (Paris)*, Vol.169, 1919, p.646; Majorana, Q., *Philosophical Magazine*, Vol.39, 1920, p.488; Majorana, Q. — *Comptes Rendues de l'Académie des Sciences (Paris)*, Vol.173, 1921, p.478; Majorana, Q., *Journal de Physique*, Vol.1, 1930, p.314; Dragoni, G. — *Proceedings of the X course on Gravitational Measurements Fundamental Metrology and Constants*, Dordrecht: Kluwer, 1988.

by a number of researchers by introducing heavy metals between the Earth and a test body. The results of these experiments suggest that there is a weakening of gravity by the intervening medium. More recently[21] it has been noticed that the artificial laser ranging satellite LAGEOS exhibits some anomalous motion whenever it is screened from the Sun's gravity by the Earth.

In an attempt to explain the flat rotation curves of the spiral galaxies and the virial mass problem in the clusters of galaxies, Milgrom[22] has proposed a revised form of Newton's second law with respect to the gravitational forces in the low-acceleration situation in the form

$$\mathbf{F} = m\mathbf{a}\left(\frac{a}{a_0}\right), \tag{2.28}$$

where a_0 is a new physical constant having the dimension of acceleration. To explain the flat rotation curves Milgrom estimated the value of a_0 to be about 2×10^{-8} cm s^{-2}. It has, of course, been shown[23] that such a modification, while solving a few difficulties, gives rise to other fresh problems. Sanders[24] and then Kuhn and Kruglyak[25] have also proposed a modification of Newton's law to resolve the flat rotation curves without invoking the hypothesis of a very large proportion of matter in the dark and invisible form. Sanders proposed a gravitational potential of the form

$$\frac{G_\infty m_1 m_2}{r}\left[1 + \alpha \exp\left(-r/r_0\right)\right], \tag{2.29}$$

where G_∞ is the value of the constant of gravitation at a very great distance between the interacting bodies. At close ranges the value of G becomes G_0, which is equal to $G_\infty(1+\alpha)$. Employing a gravitational force law derived from this potential Sanders has shown that the rotation curves for spiral galaxies become flat for a suitable choice of α and r_0. Kuhn and Kruglyak assumed the gravitational attraction to be given by

$$\frac{G_0 m_1 m_2}{r^2} + \frac{G_1 m_1 m_2}{r}. \tag{2.30}$$

They demonstrated that for a suitable choice of G_1 flat rotation curves result in spiral galaxies over a wide range of sizes. The main point to be noted in their work is the consistency of numerical results.

Before concluding I should mention one point. The above account of the research work on possible modifications of Newton's laws is far from complete. There are quite a few other published works, and the interested reader can find references to these in the research papers cited here.

---

[21] van Flandern, T., *Dark Matter, Missing Planet and New Comets*, North Atlantic Books, 1993.

[22] Milgrom, M., *The Astrophysical Journal*, Vol.270, 1983, p.365; Bekenstein, J. and Milgrom, M., *The Astrophysical Journal*, Vol.286, 1984, p.7.

[23] Felten, J.E. – *The Astrophysical Journal*, Vol.286, 1984, p.3.

[24] Sanders, R. H., *Astronomy and Astrophysics*, Vol.154, 1986, p.27.

[25] Kuhn, J. R. and Kruglyak, L., *The Astrophysical Journal*, Vol.313, 1987, p.1.

However, in most of the cases mentioned above, each of the modifications proposed stems from one particular difficulty with the standard Newton laws. None of these attempts to address all the issues simultaneously. It will be seen later that the modified form of the gravitational law, the subject matter of this monograph, has the unique distinction of being able to resolve all the problems simultaneously. Furthermore, the suggested form has quite profound implications in the field of astrophysics and cosmology.

Chapter 3

Mach's Principle and Inertial Induction

3.1 The Origin of Inertia

NEWTON'S concept of absolute space and his conjecture that inertia is an intrinsic property of matter were not accepted by some contemporary physicists and philosophers, such as Leibniz and Berkeley. However, the outstanding success of Newtonian mechanics in resolving long-standing issues and explaining the planetary and terrestrial motions overshadowed all criticisms and doubts of a philosophical nature. After about 150 years, the philosopher-scientist Ernst Mach raised a question about the absolute nature of motion. Following his predecessors of one and half century before, he advocated the relational nature of motion.

Berkeley's criticism of the idea of absolute space as conceived by Newton is presented below from his work:[1]

> But, notwithstanding what has been said, I must confess it does not appear to me that there can be any motion other than relative; so that to conceive motion there must be at least conceived two bodies, whereof the distance or position in regard to each other is varied. Hence, if there was one only body in being it could not possibly be moved. This seems evident, in that the idea I have of motion doth necessarily include relation.

The same point is reflected in his other work *De Motu* as quoted by Winkler[2]:

[1]Berkeley, G., *The Principles of Human Knowledge*, Vol.35 of *Great Books of the Western World*. Encyclopaedia Britannica, Chicago, 1952.

[2]Winkler, K. P., Berkeley, Newton and Stars: *Studies in History and Philosophy of Science*, Vol.17, 1886, p.23.

> No motion can be recognised or measured, unless through sensible things. Since then absolute space in no way affects the senses, it must necessarily be quite useless for the distinguishing of motion. Besides, determination of direction is essential to motion; but that consists in relation. Therefore, it is impossible that absolute motion should be conceived.

It appears that the phenomena associated with rotating frames led Newton to his belief in absolute space and in the absolute character of accelerations. He attempted to demonstrate this with the help of his famous bucket experiment (already mentioned in Chapter 2). In this experiment a bucket is hung by a rope and is partially filled up with water. Then the bucket is given a spin but the water (initially) remains at rest as shown in Fig.3.1a. The surface of the water remains flat. As the water gradually picks up rotation (through friction with the bucket wall), the surface becomes concave (Fig.3.1b). Next, the bucket's rotation is suddenly stopped by hand and the water continues to rotate, and the water surface remains concave. Newton showed that the curvature of the water surface resulted from the acceleration of the rotating water, and reasoned that the relative rotation of the bucket and the water was not the factor that determines the curvature of the water surface. According to him, it was the absolute rotation of the water in space and the consequent acceleration that was responsible for the phenomenon. One wonders how Newton could have been so naive as to believe that the thin wall of the bucket could have played any important role to influence the motion of the water contained. This is reflected in Mach's criticism, as he said that

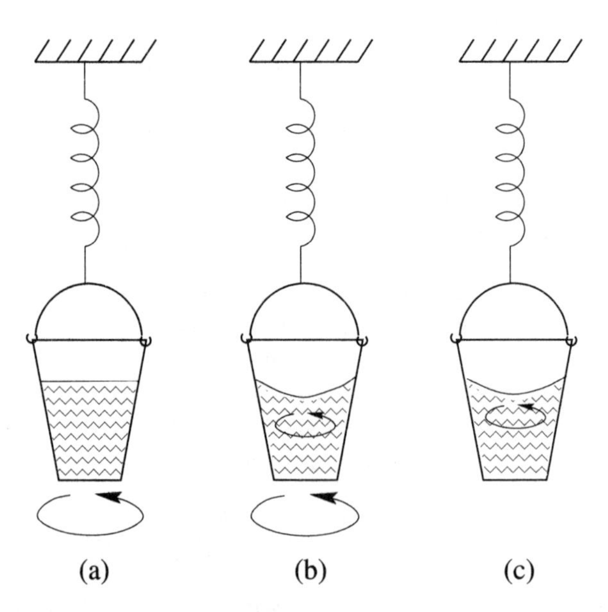

Figure 3.1: Newton's bucket experiment.

it may be only a question of degree.[3] "No one," he wrote, "is competent to say how the experiment could turn out if the sides of the vessel increased in thickness and mass until they were ultimately several leagues thick." According to Mach the motion of the water with respect to the fixed stars in the rest of the universe generated the necessary force to produce the curvature in the water surface. This is clear from the challenge he issued: "Try to fix Newton's bucket and rotate the heaven of fixed stars and then prove the absence of centrifugal forces." Mach was more explicit than Berkeley in pointing out that the stars must exert the inertial forces on an accelerating body. He wrote:

> Obviously it does not matter if we think of the Earth as turning round on its axis, or at rest while the fixed stars revolved around it. Geometrically these are exactly the same case of a relative rotation of the Earth and the fixed stars with respect to one another. But if we think of the Earth at rest and the fixed stars revolving round it, there is no flattening of the Earth, no Foucault's experiment and so on—at least according to our usual conception of the law of inertia. Now one can solve the difficulty in two ways. Either all motion is absolute, or our law of inertia is wrongly expressed. I prefer the second way. The law of inertia must be so conceived that exactly the same things result from the second supposition as from the first. By this it will be evident that in its expression, regard must be paid to the masses of the universe.

According to Mach, the inertial frame can be constituted of the very large scale structure of a universe that is infinite, homogeneous and quasistatic. Mach was more fortunate than his predecessors, and a number of prominent scientists (including Einstein) were greatly influenced by the idea and the principle known as "Mach's Principle." It has been interpreted primarily in two ways as indicated below:

1. The inertial properties of an object are determined by the presence and distribution of mass-energy throughout all space.

2. The geometry of space-time and therefore the inertial properties of every infinitesimal test particle are determined by the distribution of mass-energy throughout all space.

Einstein developed his General Theory of Relativity with the aim of incorporating Mach's Principle. Unfortunately, he did not succeed[4].

---

[3] Mach, E.– *The Science of Mechanics – A Critical and Historical Account of its Development*, Open Court, La Salle, 1960 (Originally published in 1886).

[4] For further reading on Mach's Principle refer to (i) Dicke, R. H., "The Many Faces of Mach" in: *Gravitation and Relativity*, eds. Chin, H. Y. and Hoffman, W. F., W.A. Benjamin, New York 1964, (ii) Cohen, R. S. and Seeger, R. J. (eds.), *Ernst Mach : Physicist and Philosopher*, D. Reidel Publishing Co./Dordrecht, Holland.

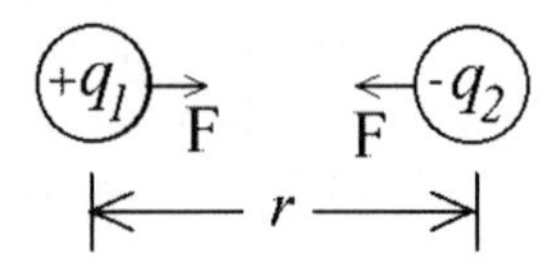

Figure 3.2: Coulomb attraction between two charged particles.

3.2 Quantifying Mach's Principle and the Concept of Inertial Induction

Though Mach's principle made a very profound impression in the minds of contemporary scientists, a quantitative description was not available. This prevented any mathematical analysis leading to quantitative results. The first major attempt in this direction was made by D. W. Sciama[5]. Noticing the similarity between Coulomb's force law for two charged particles and the inverse square law of gravitation for two particles, he proposed an acceleration-dependent term in the law of gravitation. Figure 3.2 shows two stationary charged particles with opposite charges q_1 and $-q_2$ separated by a distance r. The force with which each particle is attracted to the other is given by

$$F = \frac{\epsilon q_1 q_2}{r^2} \tag{3.1}$$

where ϵ is the dielectric constant of free space.

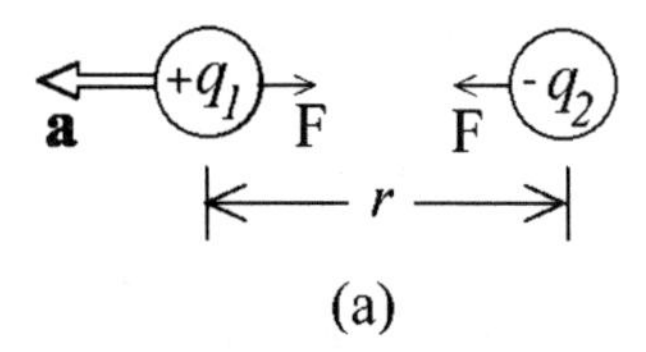

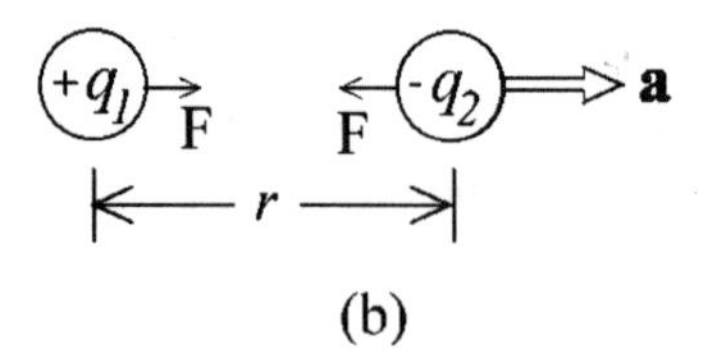

Figure 3.3: Force due to relative acceleration between charged particles.

[5]Sciama, D. W. – "On the Origin of Inertia," *Monthly Notices of the Royal Astronomical Society*, V.113, 1953, p.34.

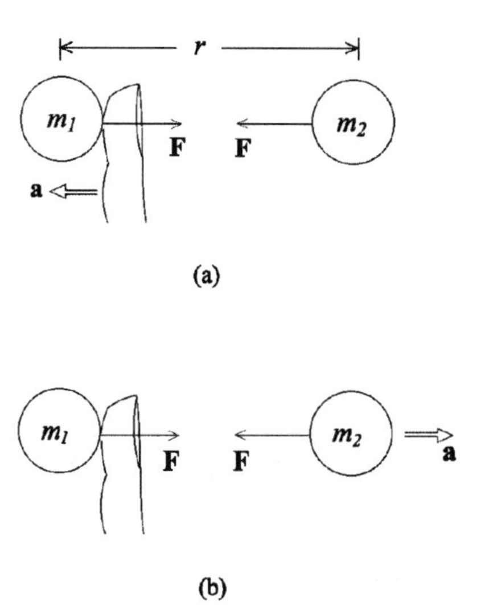

Figure 3.4: The principle of inertial induction.

Now, if the two particles possess a relative acceleration with respect to each other, the interactive force between the two has an acceleration-dependent term. For example, if particle 1 moves away from particle 2 with an acceleration a as shown in Fig.3.3 a the force F will be given by

$$F = \frac{\epsilon q_1 q_2}{r^2} + \frac{\epsilon q_1 q_2}{c^2 r} a, \tag{3.2}$$

where c is the speed of light. The situation remains the same if one assumes particle 2 to move away with respect to particle 1 as indicated in Fig.3.3b. It should be noted that the acceleration-dependent force is proportional to $1/r$ and, therefore, falls off more slowly than the static term, which falls off as the square of the distance.

Sciama considered the situation for gravitational interaction to be analogous. Thus, two particles with gravitational masses m_1 and m_2 separated by a distance r will attract each other with a force

$$F = \frac{G m_1 m_2}{r^2} + \frac{G m_1 m_2}{c^2 r} a, \tag{3.3}$$

when the relative acceleration between the two particles is a. The situation is indicated in Fig.3.4. It can be seen that as the finger pushes particle 1 away from particle 2 with an acceleration it feels the same force as when particle 2 moves away from particle 1 with the same acceleration. Sciama coined the term "inertial induction" for the acceleration-dependent extra term. Obviously, in the situation described above, **a** is

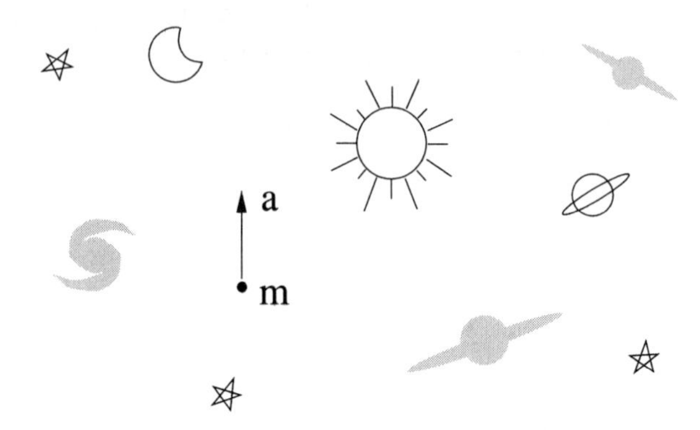

Figure 3.5: An accelerating particle in the universe.

along the line joining the particles. However, for the time being the effect of the angle which the vector **a** makes with the line joining the particles will be ignored.

Now a rough estimate of the acceleration-dependent force can be made for a particle of gravitational mass m when it is given an acceleration **a** with respect to the matter present in the rest of the universe,[6] as schematically indicated in Fig.3.5.

Of course, in so doing we assume that the stellar and galactic systems do not possess any significant systematic acceleration.[7] Since the universe can be considered to be isotropic in the large scale, the position dependent first term will get cancelled and the resultant force will be zero. (Of course, nearby heavy objects may cause a resultant pull. If we take our test particle sufficiently far away from heavy objects, then it will be almost free from any resultant gravitational pull in any particular direction.) What interests us is the resultant of the acceleration-dependent inertial induction term. The resultant force

$$F = \sum_{\text{Observable Universe}} \frac{GM_j}{c^2 r} ma, \tag{3.4}$$

where M_j is the gravitational mass of the jth object in the rest of the universe.[8] Equation (3.4) can be written in the following form as the universe may be considered to be homogeneous in the very large scale:

[6]It should be noted that the bodies present in the rest of the universe do not possess any systematic acceleration among themselves. Thus, it is possible to conceive of a frame of reference so that the systematic acceleration of the rest of the universe with respect to this frame is zero. This frame of reference is the "inertial frame of reference" in Newtonian mechanics.

[7]In fact the accelerations of the planets due to their orbital motion and those of the stars due to galactic rotation are quite small and lie in the range of 10^{-10} to 10^{-3} ms^{-2}.

[8]If the universe is infinite in its extent then of course the summation has to be done over infinite distance from the mass m.

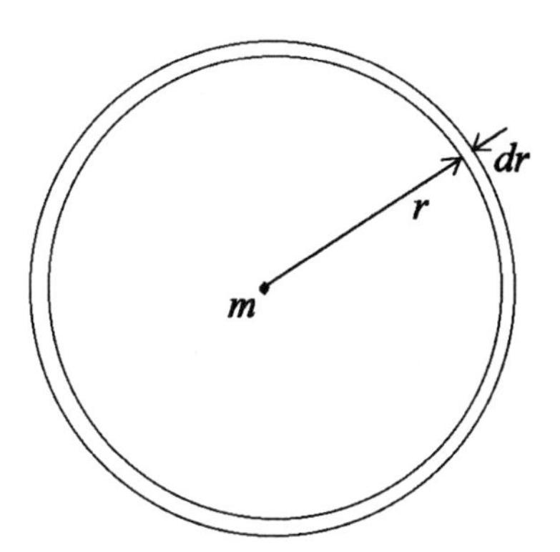

Figure 3.6: The interaction of a particle with a spherical element of the universe.

$$F = \frac{Gm\rho}{c^2}\left[\int\int_{\text{Universe}}\int \frac{dv}{r}\right] a, \tag{3.5}$$

where dv is an elemental volume of the universe and ρ is the density of gravitational mass in the universe (which is constant). The above equation can be further simplified as follows:

$$F = \frac{4\pi Gm\rho}{c^2}\int_0^{R_0=c/H} \frac{1}{r}r^2 dr \cdot a, \tag{3.6}$$

where the universe has been assumed to be composed of thin spherical shells with centre at m as indicated by one such shell shown in Fig.3.6.

The observable radius of the universe has been taken as $R_0 = c/H$ where H is the Hubble constant. It implies that the cosmological redshift is infinite, and objects become invisible at that distance. The above equation becomes

$$\begin{aligned} F &= \frac{2\pi G\rho R_0^2}{c^2} ma \\ &= \frac{2\pi G\rho}{H^2} ma. \end{aligned} \tag{3.7}$$

The average density of the universe is estimated to be approximately 10^{-26}kg m^{-3}. The estimated value of H is about 50 km s^{-1}Mpc^{-1}. Converting into SI units $H \approx 1.5 \times 10^{-18} s^{-1}$. Substituting these values in the R.H.S. of (3.7) yields the rough estimate of the total force due to inertial induction as follows:

$$
\begin{aligned}
F &\approx \frac{2 \times 3.14 \times 6.67 \times 10^{-11} \times 10^{-26}}{(1.5 \times 10^{-18})^2} ma \\
&\approx 1.8\, ma
\end{aligned}
$$

Ideally, this force should have been equal to ma, as given by Newton's second law. However, considering the approximate nature of the analysis and the uncertainties involved in the estimates of ρ and R_0, the result is quite astonishing, and it provides considerable support to Mach's hypothesis. It is therefore reasonable to accept that inertia is nothing but the manifestation of the dynamic gravitational interaction of an object with the matter present in the rest of the universe. This also resolves the mystery why the mass m in the force law (termed as the inertial mass) is equivalent to the gravitational mass. However, a serious scientific question remains unanswered. The inertial and gravitational masses are exactly equal, and not approximately equal. How is it possible for the various quantities like G, c, ρ and H to assume values which make the coefficient of ma in (3.7) exactly equal to unity?

It is interesting to note that, taking the example of Weber's electrodynamic force law (2.5), Tisserand proposed a modified law for gravitational interaction (2.6) in 1872 which contains an acceleration-dependent term very similar to the one proposed by Sciama. However, it was used to explain the advance of the perihelion of Mercury's orbit and remained unnoticed till very recently.

3.3 Relative Contributions to Inertia and Mass Anisotropy

Since the inertial effect emerges as an interactive effect with the objects present in the universe, it is appropriate to estimate the relative contributions to the inertial mass of a particle by the various objects in the universe. It has already been mentioned that the acceleration-dependent inertial induction term falls off as $1/r$. Hence, it is a long-range force, and the contribution from the huge amount of matter present in the distant universe will make a much larger contribution than the nearby less massive objects. Table 3.1 shows rough estimates of the contributions of the various objects to the inertia of a particle of mass 1 kg near the Earth's surface. Thus, it is quite clear that the primary source of inertia is the distant universe.

There is another important aspect which needs to be examined as the inertial effect is an interactive phenomenon with the matter present in the near and far regions of the universe. Since the local matter distribution is not isotropic, the contribution to inertial mass of the same object is expected to vary slightly depending on the direction of motion. Though we have not introduced any angle effect in the approximate analysis presented in the previous section, such an effect is expected to be present. Figure 3.7

Table 3.1: Contribution of various components of the universe to inertia.

Contributing object	Contribution to inertial mass of 1 kg
Earth	$\sim 10^{-8}$ kg
Sun	$\sim 10^{-7}$ kg
Milky Way galaxy	$\sim 10^{-6}$ kg
Rest of the universe	~ 1 kg

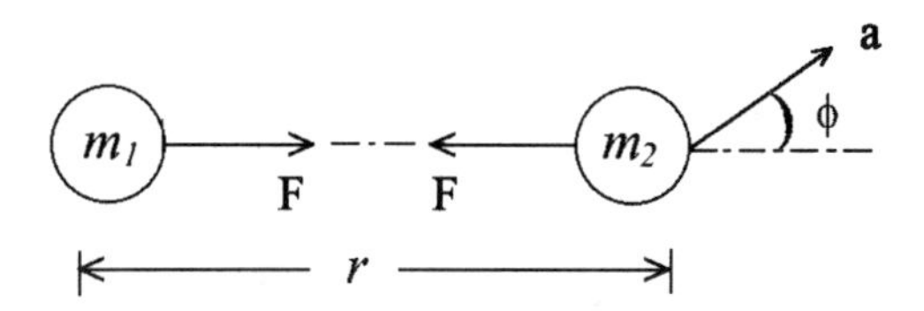

Figure 3.7: Inclination effect in inertial induction.

shows two particles in relative motion. Particle 2 has an acceleration **a** with respect to particle 1 and **a** makes an angle ϕ with the line joining the two particles, as indicated. Intuitively, it is felt that when ϕ is equal to zero the full effect of the acceleration will be felt by particle 1 due to inertial induction. On the other hand when $\phi = \pi/2$ the effect of **a** will not be felt instantaneously by particle 1. So we can write

$$F = \frac{Gm_1m_2}{r^2} + \frac{Gm_1m_2}{c^2r} af(\phi), \tag{3.8}$$

where $f(\phi)$ represents the inclination effect. Following the reasoning above we can assume the characteristics of $f(\phi)$ to be as follows:

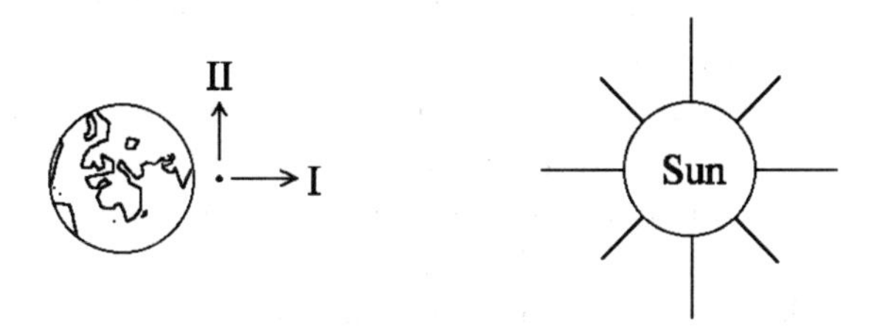

Figure 3.8: Anisotropy in inertial induction due to non-isotropy in mass distribution

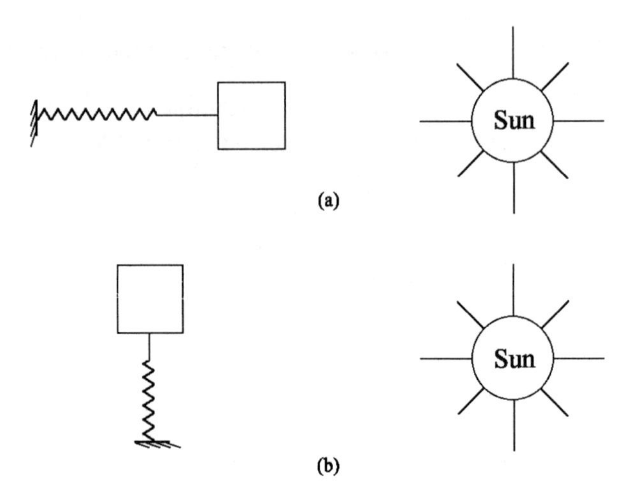

Figure 3.9: A spring mass system in different orientations with respect to the Sun.

$$
\begin{aligned}
f(\phi) &= 1 \quad \text{for} \quad \phi = 0 \\
f(\phi) &= 0 \quad \text{for} \quad \phi = \pi/2 \\
f(\phi) &= -1 \quad \text{for} \quad \phi = \pi.
\end{aligned}
\tag{3.9}
$$

Accordingly, it is reasonable to expect the inertial mass of an object, shown in Fig.3.8, to depend on the direction of acceleration. It is expected that when the particle is accelerated in direction I it will encounter greater resistance (implying larger inertia). A similar effect is also expected due to the anisotropic matter distribution resulting from the presence of our own Milky Way galaxy. However, even rigorous experiments have failed to detect any dependence of mass on the direction of acceleration. While some scientists take this negative result as an argument against Mach's principle, many others argue that this is perfectly justified, and Mach's principle is not invalidated by this. They point out that such an effect cannot be determined by taking Mach's principle only into account. In fact, the scale of force and all interactions will vary in a manner that such an effect is impossible to detect. For example, if we take a spring mass system, the frequency of natural oscillation will not depend on the situations indicated in Figs. 3.9a and 3.9b. This is so because the stiffness of the spring, which depends on the force and extension, will also change with the orientation. Force is defined as the acceleration of a unit mass, and so change in mass also causes the scale of force to change, thereby making detection impossible.

Chapter 4

Extension of Mach's Principle and Velocity-dependent Inertial Induction

4.1 Extension of Mach's Principle

MACH and other like-minded philosophers and scientists concentrated on the dependence of gravitational interaction on relative acceleration. The basic objective was, obviously, to demonstrate that inertia is nothing but the manifestation of the acceleration-dependent gravitational interaction of an object with the matter present in the rest of the universe. The problem of an interactive force depending on the relative velocity between two gravitating masses had been taken up earlier.[1] The concept of a relational mechanics which underlies Mach's principle suggests interactive gravitational forces depending on relative 'motion', and 'motion' means not only acceleration, but velocity also. It was argued by Sciama that if there did exist a velocity-dependent interaction between local matter and the distant stars, then it would produce an effect on terrestrial experiments due to our motion with respect to the stars. Such an interaction was proposed in 1963 to account for the observed decay of a particular type of elementary particle called a K-meson into two π-mesons. It was shown that the rate of the decay depends on the velocity of the K-meson relative to the stars. This

[1] Sciama, D.W. – *The Physical Foundations of General Relativity,* Heinemann Educational Books Ltd., London, 1972.

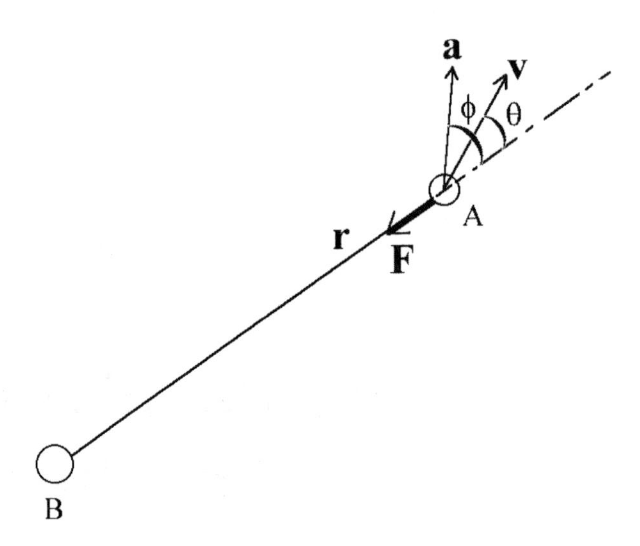

Figure 4.1: Force on a particle due to dynamic gravitational interaction.

dependence was not detected experimentally, and it was taken for granted that the dynamic gravitational interactive force does not depend on relative velocity. However, it should be pointed out that such an effect can remain undetected if it is too small. The K-meson decay experiment is capable of detecting a force up to a sensitivity of 10^{-12}. If the magnitude of the velocity dependent force is still smaller, then the experiment cannot detect it. On the other hand, one is intuitively tempted to think that such a dynamic gravitational interaction (already named "inertial induction") should oppose any relative motion.

Finally, an extension of Mach's principle is proposed to include an interactive force which depends on the relative velocity of two objects, over and above the static gravitational pull and the acceleration-dependent inertial induction effect.[2] To distinguish from the acceleration-dependent term proposed by Sciama it will be called "velocity-dependent inertial induction."

4.2 A Phenomenological Model of Dynamic Gravitational Interaction

The simplest form of a phenomenological model of dynamic gravitational interaction between two bodies can be represented as

[2]Ghosh, A.– *Pramana* (Journal of Physics), V.23, 1984, p.L671; Ghosh, A., *Pramana* (Journal of Physics), V.26, 1986, p.1.

$$\mathbf{F} = \mathbf{F_s} + \mathbf{F_v} + \mathbf{F_a} \tag{4.1}$$

where $\mathbf{F}$ is the force on A due to B (Fig.4.1), $\mathbf{F}_s$ is the static Newtonian gravitational pull, $\mathbf{F}_v$ is the force depending on the relative velocity (*i.e.*, the velocity-dependent inertial induction) and $\mathbf{F}_a$ is the force depending on the relative acceleration (*i.e.*, the acceleration-dependent inertial induction). There may be other terms depending on the higher order time derivatives of the position vector $\mathbf{r}$, but such terms will be ignored in the proposed model. Each component again may have a complicated structure. For example, the first term in $\mathbf{F}_s$ may be the usual Newtonian expression, but there may be other position-dependent terms as proposed by other researchers.[3] Likewise with $\mathbf{F}_v$ and $\mathbf{F}_a$. However, we will consider only one term in each of these. Consequently, the total force on A by B (due to dynamic gravitational interaction between them) can be expressed as follows:

$$\mathbf{F} = -\frac{Gm_A m_B}{r^2}\hat{\mathbf{u}}_r - \frac{Gm_A m_B}{c^2 r^2} v^2 f(\theta)\hat{\mathbf{u}}_r - \frac{Gm_A m_B}{c^2 r} a f(\phi)\hat{\mathbf{u}}_r, \tag{4.2}$$

where m_A and m_B are the gravitational masses [4] of A and B, respectively, v and a are the magnitudes of the relative velocity and acceleration of A with respect to B, $\hat{\mathbf{u}}_r$ is the unit vector along $\mathbf{r}$, $f(\theta)$ and $f(\phi)$ (with $\cos(\theta) = \hat{\mathbf{u}}_r.\hat{\mathbf{u}}_v$ and $\cos(\phi) = \hat{\mathbf{u}}_r.\hat{\mathbf{u}}_a$) represent the inclination effects. The exact forms of $f(\theta)$ and $f(\phi)$ are not assumed at this stage, but their forms are considered to be identical. Moreover, $f(\theta)$ is assumed to be symmetric and satisfy the following conditions:

$$\begin{aligned} f(\theta) &= \ \ 1 \quad \text{for} \quad \theta = 0, \\ f(\theta) &= -1 \quad \text{for} \quad \theta = \pi, \\ f(\theta) &= \ \ 0 \quad \text{for} \quad \theta = \pi/2. \end{aligned} \tag{4.3}$$

The logic behind these assumptions is obvious. It should be further noted that different choices of $f(\theta)$ can be made. However, we will prefer the form which yields correct quantitative results in all cases under consideration. Both $f(\theta)$ =$\cos\theta$ and $f(\theta) = \cos\theta.|\cos\theta|$ can be considered and the results will be discussed later. It should be further noted that G need not be a constant. However, it will be shown later that it decreases with distance, though to have any detectable effect the distance will have to be enormous. For all solar system and galactic problems the variation of G can be neglected. In conventional mechanics the influence of gravitation is assumed to decrease as $1/r^2$ because the flux of the agent responsible for gravitation (the gravitons) decreases as $1/r^2$. The decrease in the gravitational force is due both to the depletion

[3]For example, Milgrom, M., *Astrophysical Jr.*, V.270, 1983 p.365; Kuhn, J. R. and Kruglyak, J., *Astrophysical Jr.* V.313, 1987, p.1.

[4]Anything with energy E is treated to possess mass as per $E = mc^2$.

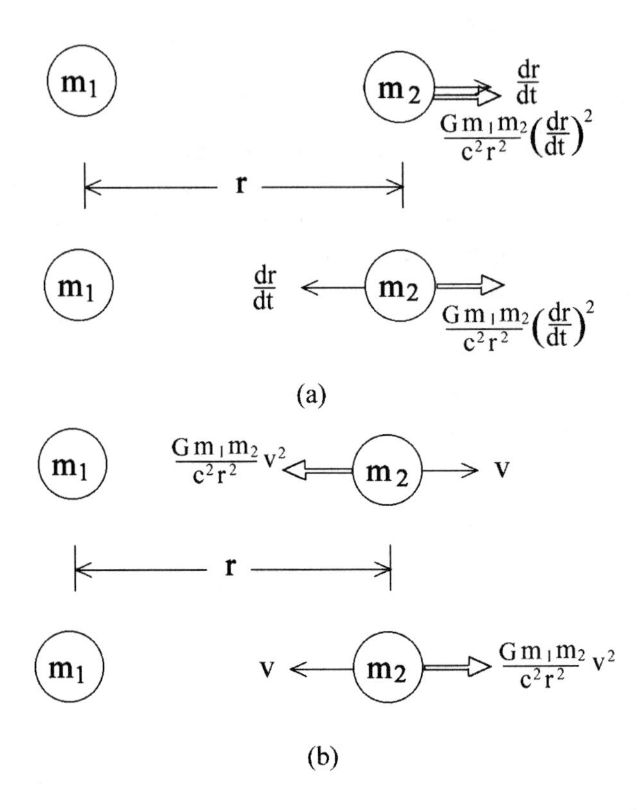

Figure 4.2: Seeliger's model and the proposed model of velocity-dependent inertial induction.

of flux as $1/r^2$ and to a decline in the strength (*i.e.*, energy) of the individual gravitons. Many researchers,[5] following Laplace, have speculated that G may decline exponentially with distance, in order to eliminate the gravitational paradox. Laplace himself had suggested the following form of gravitational force:

$$F = G_0 e^{-\lambda r} \frac{m_1 m_2}{r^2} \tag{4.4}$$

Laplace obtained the upper limit of λ from the observations of the solar system as $\lambda < 10^{-17}$ m^{-1}. It must be mentioned at this stage that it was an *ad hoc* assumption, and no value for λ could be found.

It can be seen from (2.6) that the gravitational force between two objects proposed by Tisserand also contained a velocity-dependent term. However, there is a major

[5]Laplace, P.S., "Traité de Mécanique Céleste," in: *Oeuvres de Laplace*, V.5, Book 16, Chap 4, 1880; Pechlaner, E. and Sexl, R., *Commun. Math. Phys.*, V.2, 1966, p.165; and Fuji, Y., *Gen. Rel. Grav.*, V.6, 1975, p.29.

difference between this force and the velocity-dependent inertial induction proposed here. According to our model, the velocity-dependent inertial induction term always opposes the relative velocity, as indicated in Fig.4.2b. On the other hand, as per (2.6) the direction of the velocity dependent gravitational term is independent of the direction of relative velocity as indicated in Fig.4.2a. Thus the proposed velocity-dependent inertial induction is a drag effect, whereas Tisserand's velocity term only reduces the magnitude of the static gravitational pull irrespective of the direction of the relative velocity. This is an extremely important difference and needs careful attention.

4.3 Inertial Induction and Action-at-a-Distance

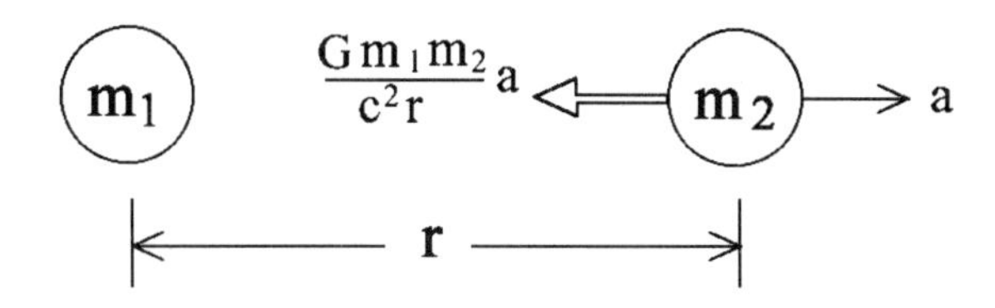

Figure 4.3: Inertial induction and action at a distance.

The concept of inertial induction suggests that an object exerts a force on another object moving with respect to the first object. This force depends on the instantaneous motion parameters. To illustrate the point let us consider Fig.4.3 which shows two particles of gravitational masses m_1 and m_2 separated by a distance r. If m_2 is given an acceleration a as shown, an inertial induction force (due to the presence of body 1) Gm_1m_2a/c^2r immediately acts on m_2 to oppose the acceleration. An objection may be raised as to how the acceleration of m_2 can be sensed by m_1 immediately, as no information can move with a speed greater than that of light. So, there should be a time lag of $2(r/c)$ between the occurrence of a and the appearance of the inertial induction force[6]. However, it should be noted that the agents carrying the effect of the object m_1 establish a field, and the interaction of this field with m_2 will be felt instantaneously by m_2. Thus any acceleration (and velocity) of m_2 will instantaneously give rise to an inertial induction effect on m_2. On the other hand, the inertial induction effect on m_1 due to any velocity and acceleration of m_2 will be felt after a period of time, r/c. It should not be considered a violation of Newton's third law, as no information can reach m_1 before this time period. This has been one of the traditional objections to Mach's principle.

[6]This is an approximate expression assuming the relative velocity between the two particles to be very small compared to c.

In the subsequent chapters the interaction of a particle with the matter present in the whole universe will be studied. Besides the effect of universal interaction, it will be shown that the local and measurable effects due to velocity dependent inertial induction also exist and can be detected.

Chapter 5

Universal Interaction and Cosmic Drag

5.1 Introduction

In the previous chapter a phenomenological model of inertial induction was proposed. The proposed form contains three terms. The first one represents the standard Newtonian gravitational force between two massive objects. The second term represents velocity-dependent inertial induction in the form of a force opposing the relative velocity. The acceleration-dependent inertial induction represented by a force opposing any relative acceleration between two objects is given by the third term. In this chapter the effect of the whole universe on the motion of a particle will be derived.

5.2 Model of the Universe

In order to derive the expression of universal interaction it is essential to have a model of the universe. It is assumed that the chosen model of the universe satisfies the perfect cosmological principle and is infinite. To satisfy the perfect cosmological principle, the universe has to be homogeneous and isotropic in the large scale. It should also remain the same (in the large scale) at all times. The universe as a whole has no overall evolution or motion, though locally, systems and objects evolve and move with finite speeds (random in nature in the large scale). Figure 5.1 shows the plot of about one million galaxies within the observable universe.[1]

[1] L. M. Lederman and D. N. Schramm, *From Quark to the Cosmos: Tools of Discovery*, Scientific American Library, 1989.

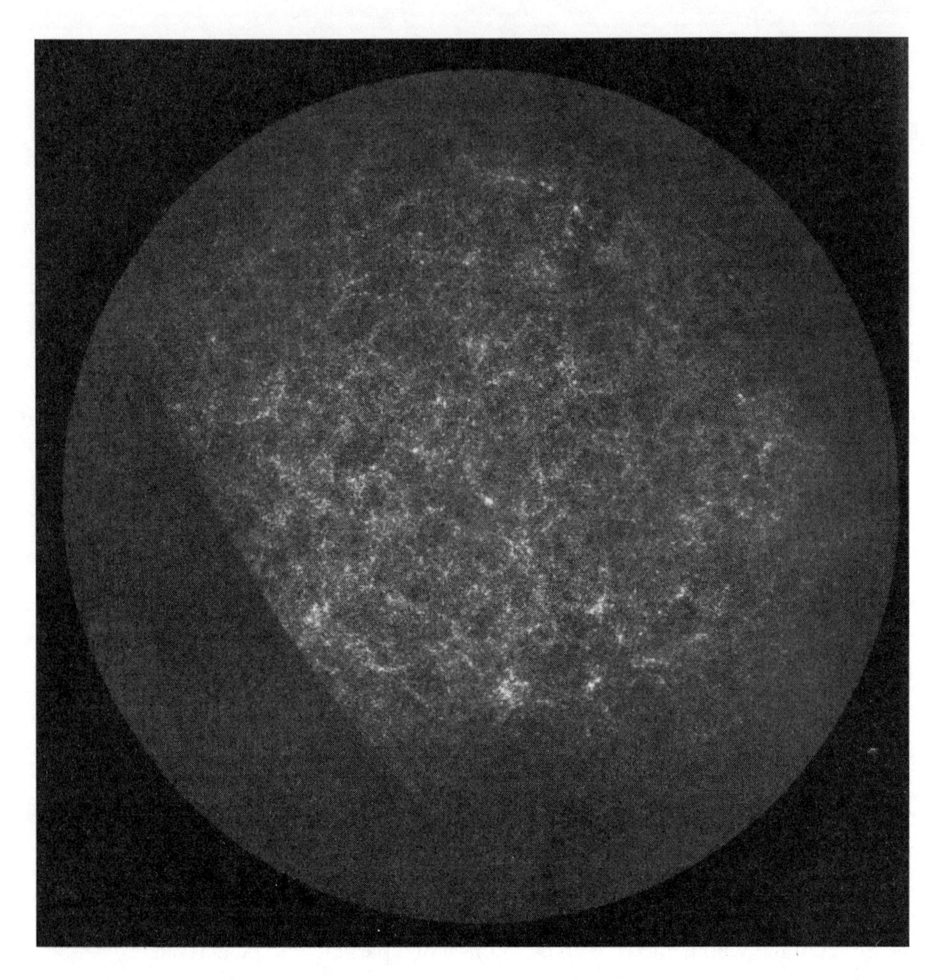

Figure 5.1: Plot of one million galaxies in the observable universe.

It can be seen from the figure that the universe is found to be homogeneous and isotropic in the large scale. After the Hubble space telescope started functioning, well formed spiral galaxies were found at very large distances. This indicates that the basic nature of the universe at a very large distance is similar to that in our vicinity. It should also be noted that the very distant objects we see now are actually pictures of these objects as they were in the distant past. Therefore, the deep sky observations indicate no universal evolution.

In fact a number of observational tests have been conducted to determine the size of the universe and to detect evidence of any universal evolution, if present. Unfortunately, none of these tests prove anything sufficiently definite, and the matter is still subject to personal bias and interpretation. However, from the philosophical point of view the perfect cosmological principle, implying an infinite quasistatic non evolving homogeneous universe, appears to be most acceptable. In the large scale, the motions of the objects present can be treated as finite and random.

To simplify the analysis we will assume the matter present in the universe to be uniformly distributed and quasistatic (*i.e.* having no systematic motion). Once the universe is considered to be infinite and quasistatic, it is possible to conceive a mean rest-frame of the universe embedded in the matter present in the universe. One way to achieve this frame is to consider three mutually separated, very large regions of the universe and join their respective centres of mass by imaginary lines. A triangle, so constructed, can represent the mean rest-frame of the universe. It was mentioned earlier that the cosmic background radiation itself constitutes such a rest-frame of the universe. When such an absolute frame of reference is available it is not meaningless to bring in the concept of absolute velocity. When we consider the universal interaction, the velocity (and acceleration also) will be measured with respect to this mean rest-frame of the universe.

5.3 Law of Motion and Cosmic Drag

The next task is to determine the interactive force between a particle and the matter present in the rest of the universe. Let us consider a particle A of gravitational mass m and let the velocity and acceleration of A with respect to the mean rest-frame of the universe be $\mathbf{v}$ and $\mathbf{a}$, respectively. The whole universe can be considered to be composed of thin concentric spherical shells with A as the centre. When integrated over the whole universe the resultant from the first term of (4.2) is zero due to the symmetry. This means that the net gravitational pull on a particle by the whole universe is zero, which is vindicated by observation and experience. In reality, the observed gravitational pull is the result of the lumpiness of the matter in the short range and its local effect. Since the theoretical calculations including only the local gravitational effects yield correct results, the absence of any universal effect is vindicated. However, the resultants of the integrated effects due to the second and third terms are not necessarily zero, as the symmetry is lost due to the vector nature of the velocity and acceleration.

To determine the resultant of the second term due to the interaction with the whole universe, let us consider an elemental ring in the thin spherical shell of radius r as shown in Fig.5.2. If the resultant force is $\delta\mathbf{F}_2$, it is obvious that it will be opposite to the velocity vector $\mathbf{v}$, and the magnitude of this velocity-dependent inertial induction of A with the elemental ring is

$$\delta F_2 = \frac{G.2\pi r^2 \sin\theta d\theta dr \rho m}{c^2 r^2} v^2 f(\theta)\cos\theta$$

where ρ is the average matter energy density of the universe.

Similarly the magnitude of the acceleration-dependent inertial induction of A with

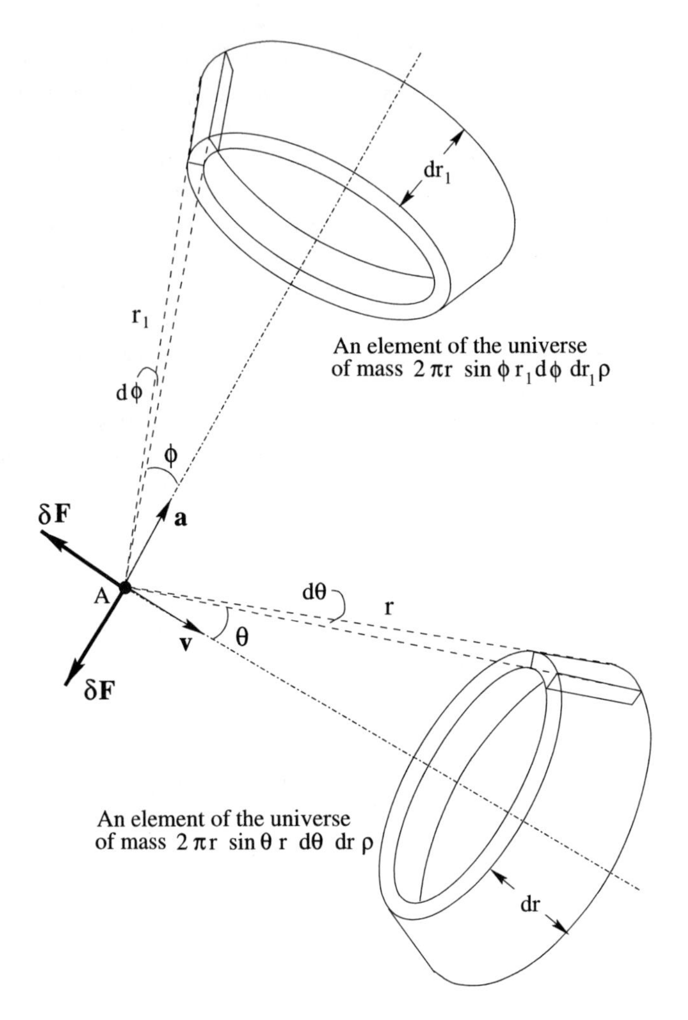

Figure 5.2: Force on a moving particle due to elements of the universe.

another elemental ring (shown in Fig.5.2) can be expressed as follows:

$$\delta F_3 = \frac{G.2\pi r_1^2 \sin\phi d\phi dr_1 \rho m}{c^2 r_1} a f(\phi)\cos\phi$$

It is also very clear that the direction of this force $\delta\mathbf{F}_3$ will be opposite to **a** due to the symmetry of the elemental ring about an axis aligned along the acceleration **a**. The reason for $\delta\mathbf{F}_2$ being opposite to **v** was also the symmetry of the elemental ring about the axis aligned with **v**.

Now the resultant force on A (moving with a velocity **v** and an acceleration **a**) due to the interaction with the matter present in the whole universe can be expressed as follows:

$$\begin{aligned}\mathbf{F} &= -2\int_0^\infty\int_0^{\pi/2} \hat{\mathbf{u}}_v \frac{G.2\pi r^2\rho\sin\theta.v^2.mf(\theta)\cos\theta d\theta dr}{c^2r^2}\\ &\quad -2\int_0^\infty\int_0^{\pi/2} \hat{\mathbf{u}}_a \frac{G.2\pi r_1^2\rho\sin\phi.a.mf(\phi)\cos\phi d\phi dr_1}{c^2r_1}\end{aligned}$$

Rewriting the above equation, we have

$$\mathbf{F} = -\hat{\mathbf{u}}_v\frac{mv^2}{c}\int_0^\infty \frac{\chi G\rho}{c}dr - \hat{\mathbf{u}}_a\frac{ma}{c^2}\int_0^\infty \chi G r_1\rho dr_1 \tag{5.1}$$

where

$$\chi = 4\pi\int_0^{\pi/2}\sin\theta\cos\theta f(\theta)d\theta = 4\pi\int_0^{\pi/2}\sin\phi\cos\phi f(\phi)d\phi \tag{5.2}$$

To get the value of χ it is necessary to have information about the forms of the function $f(\theta)$. It should be noted that the functional forms for the inclination effects of the velocity and the acceleration-dependent terms have been assumed to be identical.

To make any further progress, it is essential to know G as a function of r. The problem can be solved in the following manner. First we write

$$\int_0^\infty \frac{\chi G\rho}{c}dr = \kappa \tag{5.3}$$

Equation (5.1) representing the total force on body A can then be written as follows:

$$\mathbf{F} = -\hat{\mathbf{u}}_v\kappa\frac{mv^2}{c} - \hat{\mathbf{u}}_a\frac{ma}{c^2}\int_0^\infty \chi G r_1\rho dr_1 \tag{5.4}$$

Equation (5.4) implies that any object moving with a velocity $\mathbf{v}$ is subjected to a drag given by the first terms of (5.4). The drag is opposite to the direction of velocity and has the magnitude

$$\kappa\frac{mv^2}{c}.$$

This is a major departure from the conventional physics in which an object moving with a constant velocity is not subjected to any force. As a matter of fact, this is the basis for the first law of motion. But the concept of velocity-dependent inertial induction presupposes a force on an object moving with even a constant velocity (*i.e.* having no acceleration) in the matter-filled universe. As a result, no hadronic object can move with a constant velocity in our universe unless constantly supported by an impressed force , and the first law of motion breaks down. However, it will be seen very soon that the magnitude of the drag is so small that it cannot be detected within the scope of most present day experimental techniques.

Because of this drag, a moving object will constantly lose momentum and energy. This lost momentum and energy will go to the rest of the universe. Now we invoke the fact that gravitation is a self-acting agent. In other words, the gravitational effect can act on gravitons (agents for transporting gravitational influence) themselves. Thus the gravitons are also subjected to this universal drag and, consequently, they lose energy with distance. If a graviton has an energy E, its equivalent mass is equal to E/c^2 and the magnitude of the drag it will be subjected to is given by

$$\frac{\kappa(E/c^2)c^2}{c} = \frac{\kappa E}{c}$$

assuming that gravitons move with the speed of light. The drag opposes the velocity of the graviton. If dE is the change in energy when the graviton moves through a distance dr

$$dE = -\kappa\frac{E}{c}dr$$

(the negative sign indicates that the change is a decrease). If the energy of the graviton at the start (*i.e.* when $r = 0$) is E_0, then the above equation yields the following solution:

$$E = E_0 \exp\left[-\frac{\kappa}{c}r\right] \tag{5.5}$$

Now it should be recalled that gravitational action depends both on the flux density of gravitons and the strength (or energy) of the individual gravitons. So the gravitational influence decreases as $1/r^2$ due to the depletion of the flux in this manner and G also decreases exponentially because of (5.5). Thus we can write

$$G = G_0 \exp\left(-\frac{\kappa}{c}r\right) \tag{5.6}$$

where G_0 is the local value of the gravitational coefficient (no longer a constant), which is equal to 6.67×10^{-11} m^3 kg^{-1}s^{-2}. Substituting the above expression for G in (5.3) we obtain

$$\frac{\chi G_0 \rho}{c}\int_0^\infty \exp\left(-\frac{\kappa}{c}r\right) dr = \kappa$$

or,

$$\frac{\chi G_0 \rho}{c}.\frac{c}{\kappa} = \kappa$$

or,

$$\kappa = (\chi G_0 \rho)^{1/2} \tag{5.7}$$

Again substituting the expression for G from (5.6) into (5.4) we get

$$\begin{aligned}\mathbf{F} &= -\hat{\mathbf{u}}_v\kappa\frac{mv^2}{c} - \hat{\mathbf{u}}_a\frac{ma}{c^2}\int_0^\infty \chi\rho.G_0\exp\left(-\frac{\kappa}{c}r_1\right)r_1 dr_1 \\ &= -\hat{\mathbf{u}}_v\kappa\frac{mv^2}{c} - \hat{\mathbf{u}}_a ma.\frac{\chi G_0\rho}{\kappa^2} \end{aligned} \quad (5.8)$$

But from (5.7) $\kappa^2 = \chi G_0\rho$ and, therefore, (5.8) takes the following final form

$$\mathbf{F} = -\frac{\kappa}{c}mv^2\hat{\mathbf{u}}_v - m\mathbf{a} \quad (5.9)$$

Thus, the force acting on a particle of gravitational mass m has two Components: one depends on the velocity in the form of a drag given by the first term on the R.H.S. of (5.9). The acceleration-dependent term is exactly equal to $-m\mathbf{a}$ as given by Newton's second law.

The drag (may be called "cosmic drag") is, of course, extremely small, so that it cannot be detected by present-day experiments, though it has a very important cosmological implication which will be shown later. It is also very interesting to note that the commonly known acceleration-dependent term is identically equal to $m\mathbf{a}$ where m is the gravitational mass. This equivalence is irrespective of the magnitude of the density of the universe. Hence, no fine tuning is necessary, unlike the case of Mach's principle as quantified by Sciama. The reason for this is the self-acting nature of gravity. The attenuation of gravity by gravity itself acts as a servomechanism, and the exact equivalence of the gravitational and inertial masses is obtained. It should be further noted that such an exact equivalence will not be possible in the absence of the velocity-dependent inertial induction of the form given in (4.2).

5.4 Value of κ and the Magnitude of Cosmic Drag

A quantitative idea of the effect due to the velocity-dependent inertial induction proposed here requires a determination of the value of κ. From (5.7) we find that κ can be determined if we can complete the integrations given in (5.2) and find out χ. To satisfy the basic characteristics of the functions $f(\theta)$ representing the inclination effect, no unique choice is possible. However, it is possible to make some definite selection, as shown below.

The basic principle involved is to select $f(\theta)$ in a way that the relational characteristics of a two body system remains intact. Let us consider the case indicated in Fig.5.3. The system consists of a disc rotating at a speed Ω and a particle A at a distance r from the centre of the disc. The force on the particle due to the relative velocity of an element P of the disc will be given by

$$\delta F = \frac{Gm_1\delta m}{c^2(PA)^2}.(\Omega R)^2 f(\theta) \quad (5.10)$$

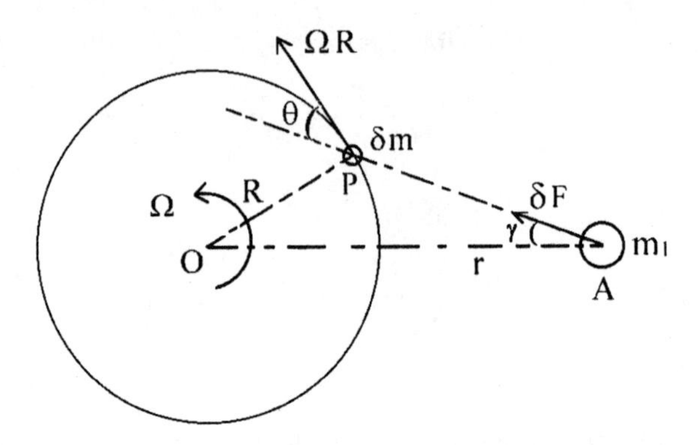

Figure 5.3: Force on a particle by a rotating body.

The criterion to select the form of $f(\theta)$ is that only the relative angular motion is important. So, if we consider the line OA to rotate with an angular velocity $-\Omega$, the force on A due to its relative velocity with respect to the element P should be the same as that given by the above expression. The situation is shown in Fig.5.4.

Thus the force δF has the following expression

$$\delta F = \frac{Gm_1 \delta m}{c^2 (PA)^2} (\Omega r)^2 f(\psi) \tag{5.11}$$

Now the force δF will be the same in both cases if

$$R^2 f(\theta) = r^2 f(\psi)$$

The above condition is satisfied if $f(\theta) = \cos\theta.|\cos\theta|$. Then

$$R^2 f(\theta) = R^2 \sin^2\beta = ON = r^2 \sin^2\gamma = r^2 \cos^2\psi$$

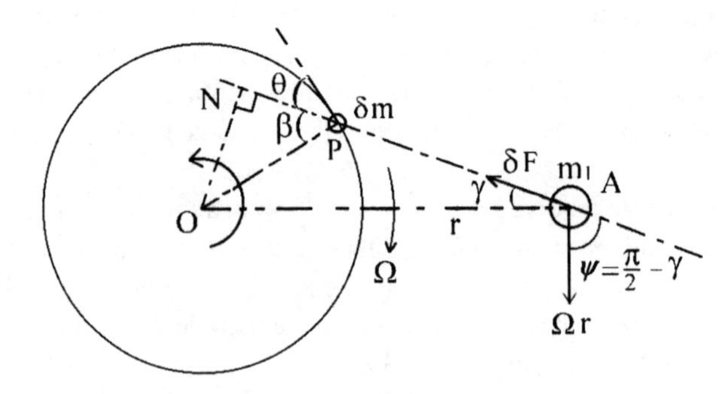

Figure 5.4: Force on a particle orbiting around a body.

Hence we will select $f(\theta) = \cos\theta.|\cos\theta|$. Using this form of the function

$$\begin{aligned}\chi &= 4\pi \int_0^{\pi/2} \sin\theta \cos^3\theta d\theta \\ &= 4\pi \int_0^{\pi/2} \sin\phi \cos^3\phi d\phi = \pi\end{aligned}$$

So the expression for κ becomes

$$\kappa = (\pi G_0 \rho)^{1/2} \tag{5.12}$$

Considering the average mass-energy density of the universe[2] equal to 7×10^{-27} kg m^{-3} and taking the standard value of G_0 as 6.67×10^{-11} m^3 kg^{-1} s^{-2}, we get

$$\kappa = 1.21 \times 10^{-18} \text{s}^{-1}$$

It is quite obvious from the above value of κ that the force due to the velocity-dependent term is extremely small and very likely cannot be detected by laboratory experiments with present-day technology. Perhaps that is why the existence of a velocity-dependent cosmic drag has not been suspected before.

The expression of the gravitational coefficient is also completely known as follows:

$$G = G_0 \exp(-\lambda r)$$

where $\lambda = \kappa/c = 0.4 \times 10^{-26}$ m^{-1}. Comparing this with the upper limit on λ as determined by Laplace ($\lambda \leq 10^{-17}$ m^{-1}) we find that it will not have any noticeable effect on the motions observed in our solar system. It is about 10 orders of magnitude smaller than the upper limit. We can also estimate the drop in the value of G across our galaxy. The diameter being about 10^5 light years, we can take $r = 10^5 \times 10^{16}$ m $\simeq 10^{21}$ m. So,

$$\begin{aligned}\frac{\Delta G}{G_0} &= \frac{G_0 - G_0 e^{-\lambda r}}{G_0} = 1 - e^{-\lambda r} \\ &\approx \lambda r \\ &\approx 0.4 \times 10^{-26} \times 10^{21} \\ &\approx 4 \times 10^{-6}\end{aligned}$$

i.e., only 0.0004% ! To get a 50% drop in the value of G we have to go very far—about 10 billion light years! Substituting this value of κ in (5.9) the force law becomes

$$\mathbf{F} = -0.4 \times 10^{-26} m v^2 \hat{\mathbf{u}}_v - m\mathbf{a}$$

where v is the velocity in m s^{-1}. Thus the cosmic drag on a one kg mass moving with a velocity of 1 m s^{-1} will be only 0.4×10^{-26} N ! An extremely small force to be measured.

---

[2]Sciama, D.W., *The Physical Foundations of General Relativity*, Heinemann Educational Books Ltd., 1972

Table 5.1: Comparison of velocity and acceleration-dependent inertial induction effects.

Interacting system	**Velocity-dependent inertial induction**	**Acceleration-dependent inertial induction**
Earth (near its surface)	$\sim 10\frac{mv^2}{c^2}$	$\sim 10^{-8}ma$
Sun (near its surface)	$\sim 275\frac{mv^2}{c^2}$	$\sim 10^{-7}ma$
Milky way galaxy (near Sun)	$\sim 200\frac{mv^2}{c^2}$	$\sim 10^{-6}ma$
Universe	$3.63 \times 10^{-10}\frac{mv^2}{c^2}$	ma

From the expressions of the velocity and acceleration-dependent terms it is clear that the acceleration-dependent inertial induction falls as $1/r$, and it is a long range force. The contribution of the matter present in the distant parts of the universe is more significant for this force. On the other hand, the velocity-dependent inertial induction force reduces as $1/r^2$, and is a relatively short-range force. Thus, the contribution of the distant matter is less significant and the cosmic drag is small compared to the acceleration dependent force. At this stage we can have some idea about the relative contributions of the inertial induction terms when we consider the interactions with the Earth, the Sun, the Milky Way galaxy and the whole universe. Table 5.1 shows the orders of magnitude of the contribution in the above mentioned cases. It is interesting to note that the magnitude of local velocity-dependent inertial induction in the vicinity of massive bodies predominates over the interaction with the rest of the universe. It must be remembered that when we considered the interaction with the whole universe, matter was assumed to be uniformly distributed. On the other-hand, in the case of acceleration-dependent inertial induction, the interaction with the whole universe is dominant. Thus, it is possible that some velocity-dependent inertial induction effects of a local nature may produce detectable results. In the later chapters a number of such phenomena will be studied.

To conclude the chapter, it is desirable to reiterate some of the very important results. It is seen that the concept of a mean rest-frame of the universe removes the ambiguity associated with the notion of inertial frames in Newtonian mechanics. The mean rest-frame of the quasistatic, infinite universe is a preferred frame of reference, and the law of motion is valid in this frame. As G has been shown to decrease exponentially with distance, the gravitational paradox is also removed. We have not only found G to be decreasing, but the rate at which it reduces is also known from the analysis. However, the most interesting result obtained from the theory is the exact equivalence of the gravitational and inertial masses. We introduce a major deviation from the conventional mechanics by supposing that all moving objects are subjected to a cosmic drag which is extremely small and almost undetectable, yet has a profound significance, as will be shown in the next chapter.

Chapter 6

Cosmic Drag : Cosmological Implications

6.1 Cosmic Drag and the First Law of Motion

Now that the stage has been set, in this chapter we shall enter into a discussion of the broader implications of cosmic drag at the cosmological level.

It should be noted that in the modified form of force law which is given by the equation (5.9) as

$$\mathbf{F} = -\frac{\kappa}{c}mv^2\hat{\mathbf{u}}_v - m\mathbf{a},$$

F represents the resistance experienced by the particle. If a particle in free space is subjected to no force, the equation of motion will be

$$m\mathbf{a} + \frac{\kappa}{c}mv^2\hat{\mathbf{u}}_v = 0,$$

or

$$\mathbf{a} = -\frac{\kappa}{c}v^2\hat{\mathbf{u}}_v.$$

Therefore **a** and **v** are along the same line but in opposite directions, implying a rectilinear motion. Writing $a = vdv/dx$ with x as the position in a rectilinear path the above equation becomes

$$\frac{dv}{dx} = -\frac{\kappa}{c}v.$$

Solving,

$$v = v_0 \exp(-\frac{\kappa}{c}x) \qquad (6.1)$$

where v_0 is the initial velocity at $x = 0$.

Equation (6.1) implies that the speed of a free particle decreases exponentially as it travels, though it does not stop. Solving the above equation and taking $x = 0$ at $t = 0$ as the initial condition

$$x = \frac{c}{\kappa} \ln \left(\frac{\kappa}{c} v_0 t + 1 \right). \tag{6.2}$$

Equation (6.2) shows that a free particle in space will travel in a straight line and it will continue to move to infinity, but with a continuously decreasing speed. The effect of the cosmic drag is similar to that of viscous drag.

6.2 Cosmic Drag and Cosmological Redshift

As has been already mentioned, the magnitude of the cosmic drag is extremely small and its effect is imperceptible in most cases. However, if we consider the fastest moving objects and the longest possible distance, the effect of the cosmic drag may show up as a detectable effect. Such objects are photons originating from the very distant galaxies. When photons lose energy, this is manifested through a decrease in the frequency. The energy of a photon, E, is given by

$$E = h\nu, \tag{6.3}$$

where h is the Planck's constant and ν is the frequency. The wavelength of a photon, λ, is given by the relation

$$\lambda\nu = c. \tag{6.4}$$

Thus, if the energy decreases by an amount ΔE, the frequency decreases by $\Delta\nu$ as follows :

$$\Delta\nu = \Delta E/h \tag{6.5}$$

This decrease in frequency is associated with an increase in the wavelength $\Delta\lambda$ so that the speed remains constant (c). Hence

$$\lambda\nu = c = (\lambda + \Delta\lambda)(\nu - \Delta\nu)$$

or,

$$\Delta\lambda \approx -\frac{\lambda}{\nu}\Delta\nu \tag{6.6}$$

taking $\Delta\lambda << \lambda$ and $\Delta\nu << \nu$.

The fractional change (increase) in wavelength $(\Delta\lambda/\lambda)$ is referred to as the redshift[1] z. Thus,

[1] Increase in the wave length causes the light to become redder, and the phenomenon is termed ‘redshift’.

$$z = \frac{\Delta\lambda}{\lambda} = -\frac{\Delta\nu}{\nu} \qquad \text{(for small } \Delta\nu\text{).}$$

Using (6.5) in the above equation

$$z = -\frac{\Delta E}{h\nu} = -\frac{\Delta E}{E}. \tag{6.7}$$

Let us now consider a photon travelling in space with an energy E. Since the equivalent mass of the photon will be given by E/c^2, the magnitude of the instantaneous cosmic drag will be given by $\frac{\kappa}{c}.\frac{E}{c^2}.c^2 = \frac{\kappa}{c}E$. When the photon travels a distance $d\xi$ the amount of energy lost dE is given by

$$dE = -\frac{\kappa}{c}Ed\xi. \tag{6.8}$$

The negative sign denotes the fact that the cosmic drag opposes the displacement $d\xi$. From (6.8)

$$\frac{dE}{E} = -\frac{\kappa}{c}d\xi,$$

and using (6.7) we get

$$\frac{d\lambda}{\lambda} = \frac{\kappa}{c}d\xi \qquad \left[\text{since } \frac{d\lambda}{\lambda} = -\frac{dE}{E}\right] \tag{6.9}$$

Solving

$$\ln\lambda = \frac{\kappa}{c}\xi + A.$$

If the wavelength at the origin (*i.e.*, at $\xi = 0$) was λ_0, then

$$A = \ln\lambda_0,$$

and we get the wavelength-distance relation as follows :

$$\ln(\lambda/\lambda_0) = \frac{\kappa}{c}x$$

where λ is the wavelength when the photon has travelled a distance x to reach us (*i.e.*, the source is at a distance x from the Earth). Rewriting the above equation

$$\frac{\lambda}{\lambda_0} = \exp\left(\frac{\kappa}{c}x\right),$$

or,

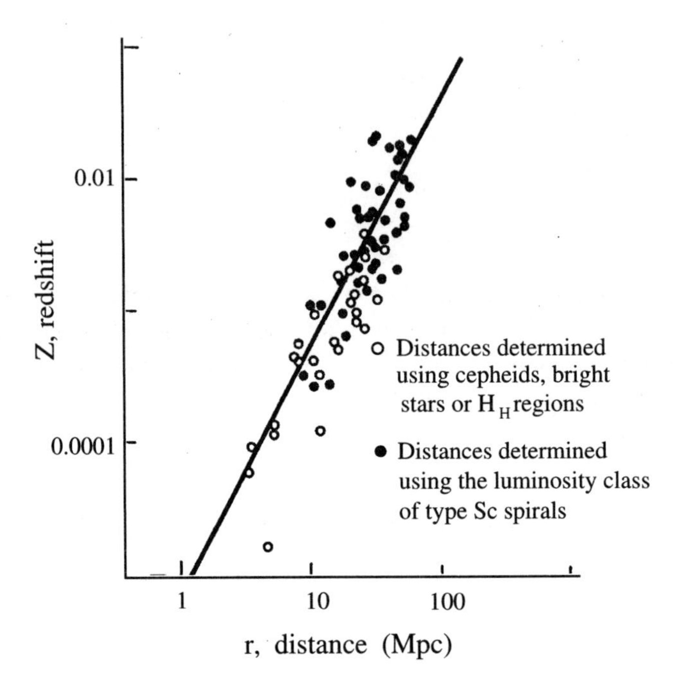

Figure 6.1: Comparison of the theoretical and observed values of cosmological redshift.

$$z = \frac{\lambda - \lambda_0}{\lambda_0} = \exp\left(\frac{\kappa}{c}x\right) - 1. \tag{6.10}$$

When $\frac{\kappa}{c}x << 1$, the above equation can be simplified (after expanding the exponential term and neglecting the higher order terms) as

$$z \approx \frac{\kappa}{c}x. \tag{6.11}$$

This means that the light from the galaxies and other distant objects will be red-shifted. Unless the distance is extremely large, the amount of redshift is proportional to the distance of the source. Since both κ and c are known, the result (6.11) can be plotted as shown in Fig.6.1. When the observational data are superimposed on this graph, the agreement with the theoretical prediction is very impressive, as is evident from the figure.

In the absence of any such cosmic drag in conventional physics, the observed red-shift has been assumed to be due to the expansion of the universe.[2] If the redshift is

[2]Usually it is explained in terms of Doppler effect caused by the recessional motion of the galaxies, but correctly speaking it is not a Doppler effect.

assumed to be due to an equivalent recessional velocity of the source, then the relation between the velocity and the fractional redshift is given by the following relation:

$$v_{rec} = cz.$$

Using (6.11) in the above equation we get

$$v_{rec} = \kappa x. \tag{6.12}$$

Thus we find that κ is nothing but the Hubble constant, H_o, which relates the recessional velocities (assumed) of the galaxies to their distances. The value of H_o obtained theoretically from (6.12), is quite close to the present estimate of 50 km s^{-1} Mpc^{-1}. It is clear from (6.10) and (6.11) that the redshift-distance relation is linear in the short range, but in the very long range it deviates from the linear relation and becomes exponential. This fact has also been supported by observations.

6.3 Hubble Anisotropy

According to the analysis presented above, the cosmological redshift is due to a velocity-dependent interaction of photons with the matter present in the universe. Thus, there is no need to invoke an hypothesis of expansion started by a Big Bang. The analysis has assumed a very ideal situation in which the matter in the universe is uniformly distributed. But in reality there is a considerable amount of inhomogeneity in matter distribution, unless we go to very large scales. Thus, in the medium range the photons arrive from different directions across space with different degrees of matter density, as indicated in Fig.6.2.

The Milky Way is shown surrounded by galaxies. The light rays coming from A and B both travel the same distance. But in case of A the path of the photons is through a region of space with higher density of matter compared to the case of B. Since the redshift is caused by the velocity-dependent inertial drag, it is expected that the magnitude of the redshift will be higher for the photons from A than for the photons from B, because of larger inertial drag for photons from A. As a result, the values of the Hubble constant[3] (which is nothing but the coefficient of proportionality in the redshift-distance relation) will be higher in the direction I than in direction II. At the same time when we go to very large distances the effect of such local density fluctuations in the universe will become less significant and the value of the Hubble constant will gradually approach the theoretical estimate. In the long range, the anisotropy in the Hubble constant should gradually vanish.

[3]Here we are referring to the coefficient of proportionality between the distance and redshift (not velocity) when we use the term Hubble constant.

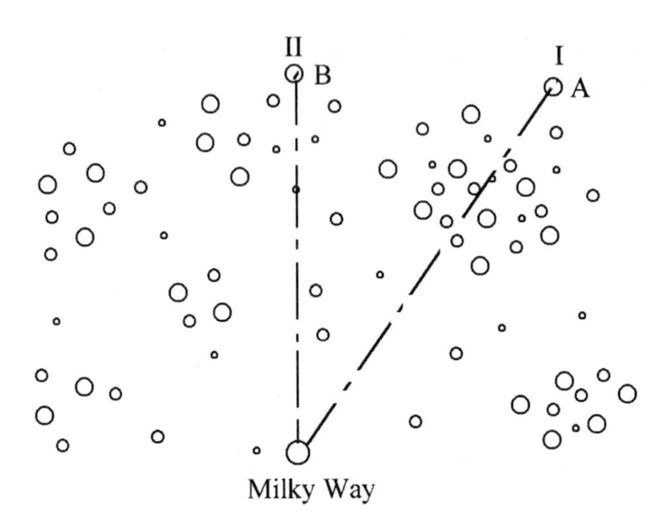

Figure 6.2: Anisotropy in the Hubble constant.

It is very interesting to note that such an anisotropy has been found to exist.[4] In 1973, Rubin *et al.* found that ScI galaxies in the magnitude range between 14 and 15 (implying all at approximately the same distance) have a mean recessional velocity[5] of 4966 $\pm$ 122 km s^{-1} in one region (I) of the sky, whereas in another region (II) the mean recessional velocity is 6431 $\pm$ 160 km s^{-1}. It has also been established that region II contains a much larger number of galaxies than region I. Subsequently other similar studies[6] also revealed the existence of an anisotropy of about 24% in the Hubble constant for the two regions. The observational data also confirm that the anisotropy effect vanishes at large distances.

6.4 Rotating Bodies in Space

Because of the cosmological drag, any rotating body should also gradually slow down. Let us consider an ideal case here. To begin with, let us take the case of a rotating ring as shown in Fig.6.3. If the ring rotates about the axis of symmetry with an angular speed Ω the cosmic drag on an element of mass δm will be given by

$$\frac{\kappa}{c}\delta m(\Omega r)^2,$$

where r is the radius of the ring. Since this drag force acts tangentially to the circular

[4]Rubin, V. C., Ford Jr., W. K., Rubin, J. S., *Astrophysical Jr. Lett.*, V.183, 1973, p.L 111.

[5]It should be noted that this recessional velocity is in fact a redshift only.

[6]Jaakkola, T., Karoji, H. *et al.*, *Mon. Not. R. Astr. Soc.* V.177, 1976, p 191; Jaakkola, T., *et al.*, *Nature* V.256, 1975, p 24.

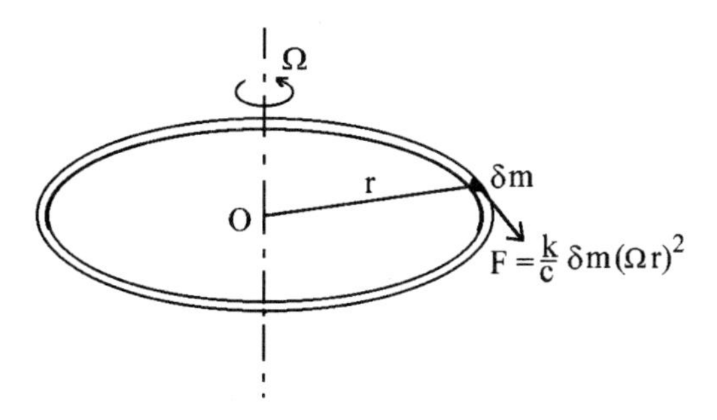

Figure 6.3: Retardation of a rotating ring in the universe.

ring, the torque developed about the centre O is

$$\frac{\kappa}{c}\delta m(\Omega r)^2 r.$$

Integrating for all the elements, the total torque resisting the rotation can be written as

$$\frac{\kappa}{c}\Omega^2 r^3 m,$$

where m is the total mass of the uniform ring. Therefore, the resulting angular deceleration can be found by dividing this resistive torque by the moment of inertia mr^2. Thus

$$\dot{\Omega} = -\frac{\kappa}{c}\Omega^2 r.$$

Solving this we get

$$\Omega = \frac{\Omega_0}{1 + \frac{\kappa}{c} r \Omega_0 t},$$

where Ω_0 is the initial angular speed at $t = 0$. It is seen that as t increases, Ω decreases but the rate of decrease of speed depends on the radius.

It has been demonstrated in the above sections that though the effect of the proposed velocity-dependent inertial induction term is very small, it leads to some consequences of a profound nature. We find that cosmic drag always slows down moving objects in free space and, as a result, the first law of motion is not strictly relevant in our universe over long distances. Furthermore, the universe is shown to be non-expanding and the Origin of the cosmological redshift is found to be interactive in nature. A few more interesting results of the proposed theory will be presented in the subsequent chapters.

Chapter 7

Velocity-dependent Inertial Induction : Local Interaction of Photons with Matter

7.1 Introduction

IT has been already mentioned that so far as the velocity-dependent inertial induction is concerned, local interactions produce much larger effects than the universal interaction without the presence of local massive objects. It is also possible that many such effects may be detectable. Therefore, it is desirable to study situations that could result in observable effects due to inertial induction of a local nature. There can be primarily two types of such interaction: (i) photons with matter, and, (ii) matter with matter. At first we shall take up a few cases of velocity-dependent inertial induction between photons and matter.

7.2 Gravitational Redshift on the Surface of a Planet (or a Star)

In the conventional mechanics it can be shown that photons lose energy (and become redshifted) when moving against gravitation. Figure 7.1 shows a photon moving away from a gravitational field. If the planet (or star) on whose surface the phenomenon is taking place has a mass M and radius R the gravitational pull on the photon of energy E will be given by[1]

[1] Variation in E is neglected, assuming it to be small.

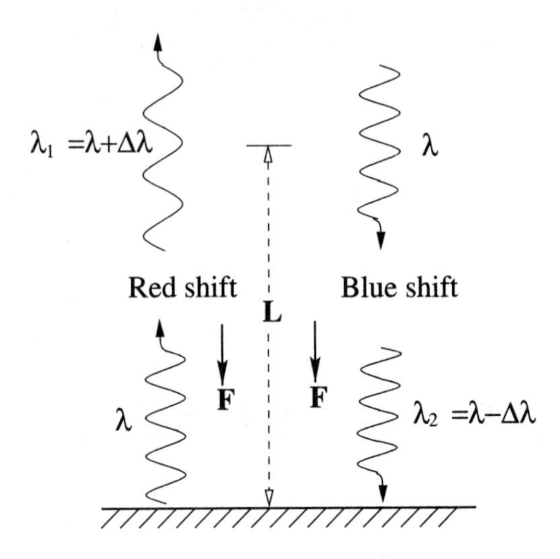

Figure 7.1: Rising and falling photons on the Earth surface.

$$F = \frac{GME}{R^2c^2},$$

so long as $L << R$, where L is the height of the photon from the surface of the gravitating body. The energy of the photon when it reaches a distance L will be equal to $E - FL$. Thus the new wavelength λ_1 is given by

$$\lambda_1 = \frac{hc}{E - FL} = \lambda + \Delta\lambda, \quad (7.1)$$

where λ is the original wavelength on the surface and $\Delta\lambda$ is the amount of increase in wavelength. Again, we know that

$$\lambda = \frac{hc}{E}. \quad (7.2)$$

Using (7.1) and (7.2)

$$\frac{hc}{E}\left(1 + \frac{\Delta\lambda}{\lambda}\right) = \frac{hc}{E - FL},$$

or,

$$1 + \frac{\Delta\lambda}{\lambda} = \frac{1}{1 - \frac{FL}{E}}. \quad (7.3)$$

If $FL << E$, *i.e.*, when L is not too large,

$$\frac{1}{1-\frac{FL}{E}} = 1 + \frac{FL}{E} - \dots$$

Using this in (7.3) and neglecting the higher order terms in (FL/E) we get

$$\begin{aligned} \frac{\Delta\lambda}{\lambda} &\approx \frac{FL}{E} \\ &\approx \frac{GM}{R^2c^2} \cdot L. \end{aligned} \tag{7.4}$$

Similarly when a photon of wavelength λ comes from a distance L to the surface of the planet, it gains energy, and hence is subjected to a blueshift, as shown in Fig. 7.1. By an analysis similar to the above it can be shown that the relative blue shift is as follows:

$$\frac{\Delta\lambda}{\lambda} \approx \frac{GM}{R^2c^2}L,$$

where the final wavelength of the photon

$$\lambda_2 = \lambda - \Delta\lambda. \tag{7.5}$$

If λ_1 and λ_2 can be measured then

$$\lambda_1 - \lambda_2 = 2\Delta\lambda,$$

or,

$$\Delta\lambda = \frac{\lambda_1 - \lambda_2}{2}. \tag{7.6}$$

An experiment of this type was successfully conducted by Pound and Rebka, (and later by Vessot *et al.*) [2] on the surface of the Earth. Pound and Rebka found the fractional frequency shift to be about -19.7×10^{-15} with the source on the surface of the Earth and the receiver at a distance of 72 feet, and -15.5×10^{-15} with the source at the top and the receiver at the bottom. It is obvious that a considerable amount of frequency shift is present due to other reasons. If we consider this extra amount, ϵ, to be same for both direction, then $\lambda_1 = \lambda + \Delta\lambda + \epsilon$ and $\lambda_2 = \lambda - \Delta\lambda + \epsilon$. Thus, the fractional redshifts for the upwards and downwards directions will be $(\Delta\lambda + \epsilon)/\lambda$ and $(\epsilon - \Delta\lambda)/\lambda$, respectively. The difference of these two fractional shifts divided by two yields the desired fractional redshift $(\Delta\lambda/\lambda)$. Using these results, we find the fractional redshifts to be 2×10^{-15} as predicted by the calculations.

[2]Pound, R. V. and Rebka, G. A., *Phys. Rev. Lett.*, V.4, 1960, p. 337; Vessot, R. F. C. *et al.*, *Phys. Rev. Lett.*, V.45, 1980, p. 2081.

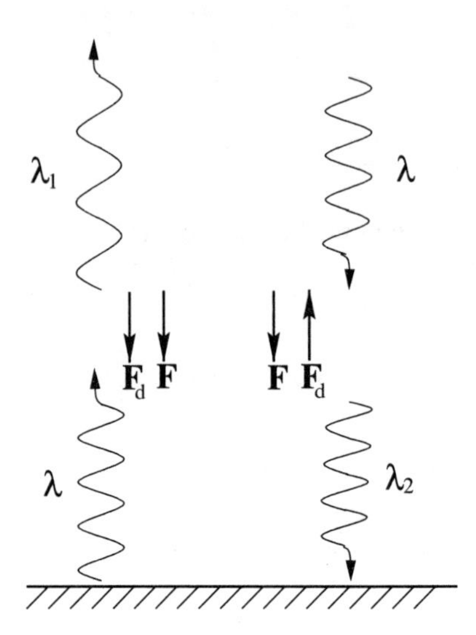

Figure 7.2: Force on falling and rising photons due to inertial induction.

According to the theory proposed here, two forces act on a photon on the surface of the planet—the gravitational pull (F) and the velocity dependent inertial drag (F_d) as indicated in Fig.7.2. However, F always acts downwards towards the surface, whereas the velocity-dependent inertial drag F_d opposes any motion. Consequently,

$$\frac{\lambda_1 - \lambda}{\lambda} \approx (F + F_d)L/E$$

gives the change of wave length when the photon travels upwards. For the falling photon

$$\frac{\lambda_2 - \lambda}{\lambda} \approx (-F + F_d)L/E$$

Thus

$$\lambda_1 - \lambda_2 \approx 2\frac{FL}{E}\lambda$$

and the effect of the drag term is cancelled. This indicates that the usual two-way experiments of the type conducted by the researchers will not reveal the existence of the velocity-dependent inertial induction. Pound and Snider [3] mentioned some 'one-way' experiments, but the detailed results are not available. However, the magnitude of the frequency shifts for each one-way reading suggests that these can contain the effect due to the velocity-dependent inertial induction also.

[3]Pound, R. V. and Snider, J. L., *Phys. Rev.* V.B 140, 1965, p. 788.

7.3 Redshifts in White Dwarfs

It is seen from the previous section that the existence of the velocity-dependent inertial induction can be detected by studying the redshift in the light coming from massive stars. The fractional shift in frequency (or, fractional redshift) of photons coming from a star of mass M and radius r is given by

$$\frac{G_0 M}{c^2 r}$$

If we know the mass and radius of a star and can measure the fractional redshift, it will be possible to find out if any excess red shift is present due to the velocity-dependent inertial drag. However, the photospheres of normal stars have a considerable amounts of turbulence and gravitational effects. For this reason, the luminous gas possesses a large proportion of radially outward velocity. This influences the amount of redshift, the resultant shift in frequency being due to both gravitational pull, velocity drag and Doppler effect because of the source's motion. In the next section we shall take up the case of redshift in the solar spectrum.

White dwarfs are degenerate stars and have much higher densities than normal stars. In normal stars the fractional redshift of emitted light is of the order of 10^{-7}, whereas in the case of white dwarfs it is about 10^{-4}. Moreover, the density being much larger, the granulation phenomenon is not present in these stars and no Doppler effect contaminates the redshift of the emitted photons. Hence, if the proposed velocity-dependent inertial drag is present, then the total intrinsic fractional redshift should be greater than $(G_0 M/c^2 r)$. Conversely, if we calculate the mass M from the measured intrinsic redshift without allowance for the effect of the extra drag term, it should be found to be substantially higher than the true value.

If we consider the velocity-dependent inertial induction of a photon leaving a star (Fig.7.3), the velocity drag is given approximately by

$$F_d \approx \frac{G_0 M m_p}{\xi^2} \cos^3 \phi.$$

where m_p is the equivalent mass of the photon,[4] ξ and θ are shown in the figure and can be expressed as follows :

$$\xi = \left[r^2 \sin^2 \theta + (r \cos \theta + x)^2 \right]^{1/2}$$

and

$$\phi = \cos^{-1} \left[\frac{r \cos \theta + x}{\left\{ r^2 \sin^2 \theta + (r \cos \theta + x)^2 \right\}^{1/2}} \right].$$

[4]Since the fractional frequency shift is small we may take m_p to be constant. We also treat the mass of the star as concentrated at the centre, which is not exactly true; but the effect of this assumption on the result will not be very large.

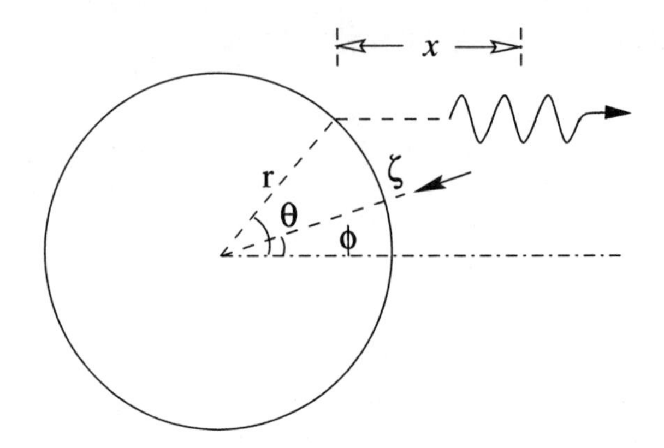

Figure 7.3: Force on a photon moving away from an object.

r being the radius of the star and θ the latitude of the emission point with respect to the plane passing through the observer. The loss of energy when the photon travels a small distance dx is given by

$$dE = -\frac{G_0 M m_p (r\cos\theta + x)^3}{\left[r^2\sin^2\theta + (r\cos\theta + x)^2\right]^{5/2}} dx, \tag{7.7}$$

or,

$$hd\nu = -\frac{G_0 M h\nu (r\cos\theta + x)^3}{c^2\left[r^2\sin^2\theta + (r\cos\theta + x)^2\right]^{5/2}} dx. \tag{7.8}$$

Solving the above differential equation

$$\ln\left(\frac{\nu + \Delta\nu}{\nu}\right) = -\frac{G_0 M}{c^2 r}\left(1 - \frac{1}{3}\sin^2\theta\right) \tag{7.9}$$

Since $\Delta\nu/\nu << 1$, higher order terms in $(\Delta\nu/\nu)$ can be ignored and the above equation can be written in the simplified approximate form as follows:

$$\frac{\Delta\nu}{\nu} \approx -\frac{G_0 M}{c^2 r}(1 - \frac{1}{3}\sin^2\theta), \tag{7.10}$$

or,

$$\frac{\Delta\lambda}{\lambda} \approx \frac{G_0 M}{c^2 r}(1 - \frac{1}{3}\sin^2\theta). \tag{7.11}$$

Thus we see that the frequency shift (or, redshift) depends on the location from which the photon is omitted. When $\theta = 0$,

$$\frac{\Delta\lambda}{\lambda} \approx \frac{G_0 M}{c^2 r}. \tag{7.12}$$

which is same as that for the redshift due to gravitational pull. When $\theta = \pi/2$

$$\frac{\Delta\lambda}{\lambda} \approx \frac{2G_0M}{3c^2r}. \tag{7.13}$$

For the overall effect we may take the average value of the redshift

$$\frac{\Delta\lambda}{\lambda} \sim 0.67\frac{G_0M}{c^2r}. \tag{7.14}$$

The fractional redshift of a photon due to the Newtonian gravitational pull is given exactly by

$$\frac{\Delta\lambda}{\lambda} = \frac{G_0M}{c^2r} \tag{7.15}$$

irrespective of the value of θ. This can be taken as the fractional redshift of the light from the whole star due to conventional gravitational pull. Finally the total fractional redshift of light from a white dwarf star can be expressed as follows :

$$z = \frac{\Delta\lambda}{\lambda} = \frac{G_0M}{c^2r} \quad \text{(without velocity-dependent inertial drag),} \tag{7.16}$$

and

$$z = \frac{\Delta\lambda}{\lambda} = 1.67\frac{G_0M}{c^2r} \quad \text{(with velocity-dependent inertial drag).} \tag{7.17}$$

Thus, if we measure the redshift, the relativistic (or gravitational) mass of the star can be expressed as follows for the above two situations:

$$M(\text{conv}) \sim \frac{zc^2r}{G_0}, \tag{7.18}$$

$$M(\text{proposed}) \sim \frac{zc^2r}{1.67G_0}. \tag{7.19}$$

However, the mass of a white dwarf can be determined by other methods, and the value obtained by such methods is termed as "astrophysical mass" and denoted by M_a. Comparison of the results for M from the above two relations with the values of M_a for a number of stars would give us an insight as to which one of (7.18) and (7.19) is a better picture of reality. Table 7.1 shows the results compiled by Shipman and Sass (1980) along with the value of relativistic mass obtained using (7.18) and (7.19).

Since 1967 it has been noted by a number of astronomers[5] that the "relativistic" (or, gravitational) masses (without considering velocity-dependent inertial drag effects) of white dwarf stars are significantly greater than the "astrophysical" masses.

[5]Greenstein, J. L. and Trimble, V. L., *Astrophysical Jr.*, V.149, 1967, p. 283; Moffett, T. J.,*et al.*, *Astronomical Jr.*, V.83, 1978, p. 820; Shipman, H. L. and Sass, C.A., *Astrophysical Jr.*, V.235, 1980, p. 177; Grabowski, B. *et al.*, *Astrophysical Jr.*, V.313, 1987, p. 750.

Table 7.1: Mean mass of white dwarf stars by various methods.

Method	No.of Stars	Mean Mass
Photometry	110	$0.55M_{\odot}$
Photometry	31	$0.60M_{\odot}$
Binary Stars	7	$0.73M_{\odot}$
Two-colour diagram	40	$0.60M_{\odot}$
Two-colour diagram	35	$0.45M_{\odot}$
H-line profiles	17	$0.55M_{\odot}$
Combined	240	Average $M_a \approx 0.55M_{\odot}$
Gravitational redshift (conventional)	83	Average $M \approx 0.80M_{\odot}$
Redshift (proposed)	83	Average $M \approx 0.50M_{\odot}$

Where $M_{\odot}$ is the mass of the Sun.

Various attempts to explain the discrepancy have been made, but the problem is still unsolved. However, if (7.19) is used to determine the gravitational masses instead of the conventional equation (7.18), then the discrepancy disappears.

As the values of M_a and M were expected to be same, a prodigious effort has been made by researchers to bring M_a and M into closer agreement. However, more data (without any bias) obtained using recent advanced techniques may throw more light on the subject.

7.4 Excess Redshift of Solar Spectrum at the Limb

Though our information about distant stars and white dwarfs is not precise, we know all the relevant data so far as the Sun is concerned. Therefore, let us attempt to determine the redshift in the light coming from the Sun. To make the results convincing, we will try an accurate calculation rather than a rough approximation.

Figure 7.4 shows the Sun and a photon at the location B at a given instant. The photon was emitted at B_0 on the Sun's surface. The photon of mass m_p is moving with a speed c in the x-direction as indicated. Let us now consider an elemental volume of the Sun at A (a location ξ, ϕ, ψ in the spherical co-ordinate system). The mass of the elemental volume is dM. If the density of the Sun is given by a function $\rho(\xi)$, we have

$$dM = \rho(\xi)\xi^2 d\phi d\psi d\xi. \tag{7.20}$$

If **dF** be the force on the photon due to the velocity-dependent inertial induction of

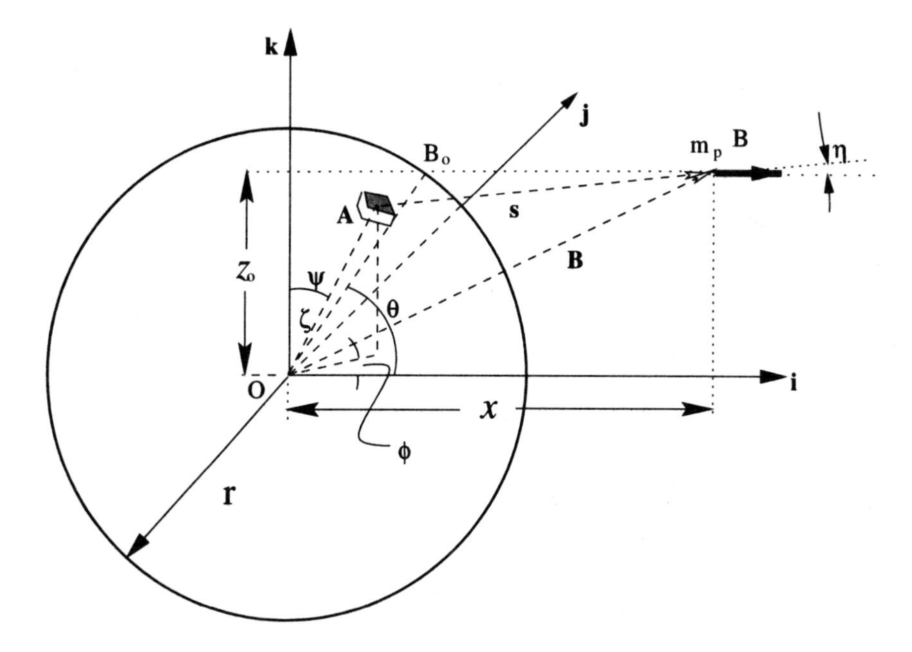

Figure 7.4: Force on a photon moving away from the Sun.

the elemental mass dM, then

$$d\mathbf{F} = -\frac{G_0 m_p dM}{c^2 s^2}.c^2 \cos\eta|\cos\eta|\hat{\mathbf{s}},$$

where $\hat{\mathbf{s}}$ is the unit vector along the vector **s** denoting the instantaneous position of the photon with respect to the elemental mass dM. The angle between the vector s and $\hat{s}$ is η, or,

$$\cos\eta = \hat{\mathbf{s}}.\hat{\mathbf{i}}.$$

The component of the elemental force $d\mathbf{F}$ along the x-direction

$$\begin{aligned} dF_x &= -\frac{G_0 m_p dM}{c^2 s^2}.c^2 \cos\eta.|\cos\eta|\left(\hat{\mathbf{s}}.\hat{\mathbf{i}}\right) \\ &= -\frac{G_0 m_p dM}{s^2}\cos^2\eta.|\cos\eta|. \end{aligned} \tag{7.21}$$

Again from Fig.7.4, we know the following :

$$\mathbf{s} = \mathbf{B} - \mathbf{A} \tag{7.22}$$

where **B** represents the position vector of the photon and **A** represents the position vector of the elemental mass at A. Thus

$$\mathbf{B} = \hat{\mathbf{i}}x + \hat{\mathbf{k}}z_0. \tag{7.23}$$

$$\mathbf{A} = \hat{\mathbf{i}}\xi \sin\psi \cos\phi + \hat{\mathbf{j}}\xi \sin\psi \sin\phi + \hat{\mathbf{k}} \cos\psi. \tag{7.24}$$

We also know that

$$\cos\eta = \frac{1}{s}\left(x - \xi \sin\psi \cos\phi\right) \tag{7.25}$$

and

$$s^2 = (x - \xi \sin\psi \cos\phi)^2 + \xi^2 \sin^2\psi \sin^2\phi + (z_0 - \xi \cos\psi)^2. \tag{7.26}$$

Using (7.20) and (7.25) in (7.21) we get

$$dF_x = -\frac{G_0 m_p \left[\rho(\xi)\xi^2 \sin\psi |x - \xi \sin\psi \cos\phi| \cdot (x - \xi \sin\psi \cos\phi)^2\right]}{s^5} d\xi d\psi d\phi \tag{7.27}$$

The x-component of the total force on the photon is given by

$$F_x = \int_{\text{solar volume}} dF_x$$

The amount of energy lost by the photon in travelling from point B_0 to the Earth (assumed to be at an infinitely large distance without introducing any perceptible change in the result) can be expressed as follows :

$$\Delta E = \int_{x_o}^{\infty} dx \int_{\text{solar volume}} dF_x \tag{7.28}$$

where $x_0 = r\cos\theta$. Using (7.27) in (7.28), and integrating over the solar volume we obtain[6]

$$\Delta E = -G_0 m_p \int_{r\cos\theta}^{\infty} dx \int_0^r d\xi \int_0^{\pi} d\psi \int_0^{2\pi} d\phi \left\{ \frac{\rho(\xi)\xi^2 \sin\psi |x - \xi \sin\psi \cos\phi| (x - \xi \sin\psi \cos\phi)^2}{s^5} \right\} \tag{7.29}$$

[6]Since the change in energy of the photon is much less than its energy, any variation in $m_p (= h\nu)$ can be ignored and m_p can be assumed to be constant while integrating along x.

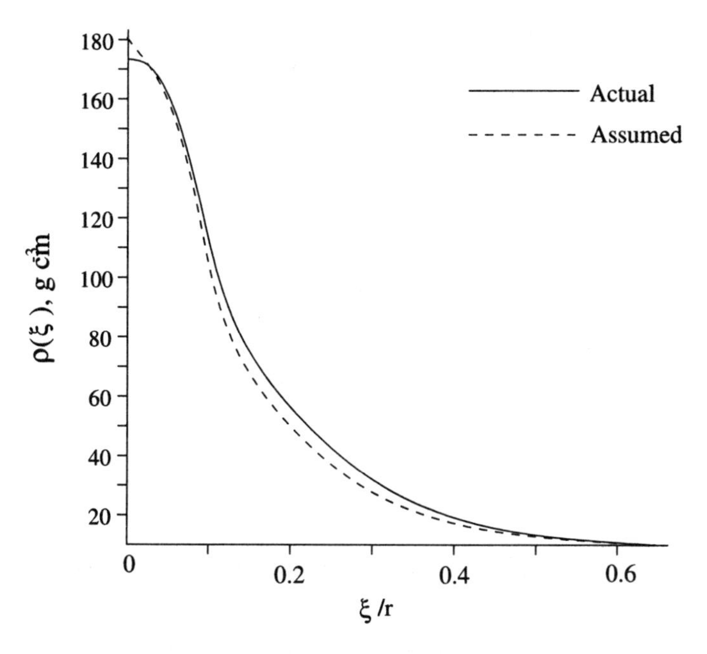

Figure 7.5: Density distribution in the Sun.

In order to carry out the integration it is necessary to know the density function $\rho(\xi)$. Figure 7.5 shows the variation of density in the Sun. The actual density function can be expressed approximately as

$$\rho(\xi) = 23 \times 10^4 \exp(-10\xi/r) \quad \text{kg m}^{-3} \tag{7.30}$$

Now, the loss of energy by the photon arriving at the Earth will show up as an increase of the wavelength (or in a decrease in frequency); if $\Delta\nu_{\text{drag}}$ is the decrease in frequency, then

$$\Delta E \approx h\Delta\nu_{\text{drag}} \tag{7.31}$$

as shown before. Using (7.26) and the density function in (7.29) we get, after taking note of (7.31) and integrating over x

$$\begin{aligned} \frac{\Delta\nu_{drag}}{\nu} \approx & -\frac{2G_0 \times 23 \times 10^4}{c^2} \int_0^r d\xi \int_0^\pi d\psi \int_0^\pi d\phi e^{-\frac{10\xi}{r}} \\ & \times \left[\frac{1}{\{r^2 + \xi^2 - 2r\xi(\sin\theta\cos\psi - \cos\theta\sin\psi\sin\phi)\}^{1/2}} \right. \end{aligned}$$

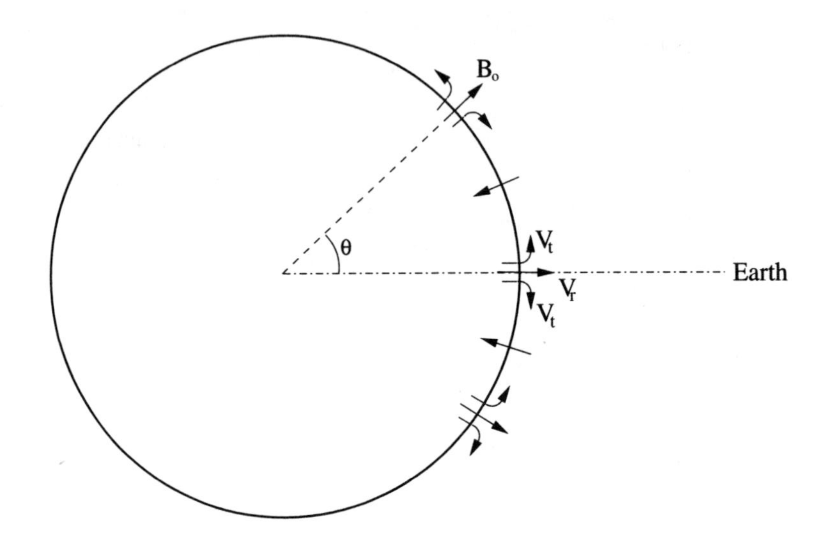

Figure 7.6: Material flow due to solar granulation effect.

$$+ \quad \frac{\xi^2 \sin^2\psi \sin^2\theta - \xi^2 - r^2\sin^2\theta + 2\xi r\cos\psi\sin\theta}{3\left\{r^2+\xi^2-2r\xi(\sin\theta\cos\psi-\cos\theta\sin\psi\sin\phi)\right\}^{3/2}}\Bigg] \tag{7.32}$$

The above equation can be rewritten in the following form:

$$\frac{\Delta\nu_{drag}}{\nu} \approx \frac{G_0 M_\odot}{c^2 r}\left\{\frac{r}{M_\odot} I(\theta)\right\}, \tag{7.33}$$

where $M_\odot$ is the solar mass and

$$I(\theta) = 46 \times 10^4 \int_0^r d\xi \int_0^\pi d\psi \int_0^{2\pi} d\phi e^{-10\xi/r}[\ ],$$

where the term [ ] is the same as that within square bracket in the R.H.S. of (7.32).

It should be noted that the above frequency shift is due to the velocity-dependent inertial induction term only. Over and above this, the Newtonian gravitational pull will also produce a frequency shift given by

$$\frac{\Delta\nu_{grav}}{\nu} = \frac{G_0 M_\odot}{c^2 r},$$

and this is independent of the location from which the photon is emitted. Therefore, the total frequency shift can be expressed as follows:

$$\frac{\Delta\nu}{\nu} \approx \frac{G_0 M_{odot}}{c^2 r}\left\{1 + \frac{r}{M_\odot} I(\theta)\right\} \tag{7.34}$$

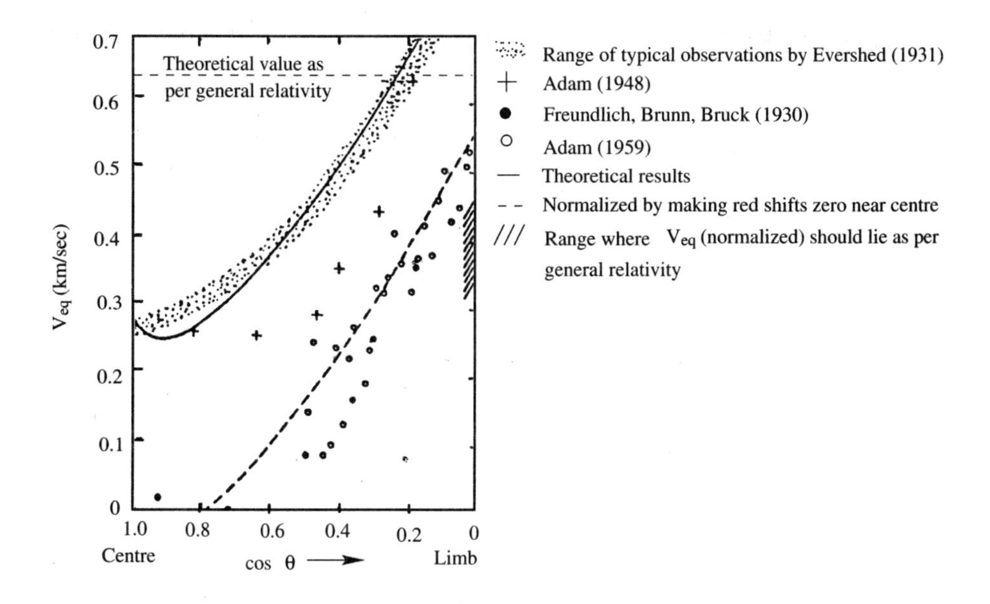

Figure 7.7: Redshift in the solar spectrum—theoretical prediction and observational results.

The resultant redshift (or frequency shift), after subtracting other redshifts due to Earth's orbital motion and Sun's rotation, can be represented by an "equivalent velocity of recession" which can produce the same redshift by Doppler effect. The magnitude of this equivalent recessional velocity corresponding to the term $G_0 M_\odot / c^2 r$, is equal to 0.636 km/s. However, the material in the photosphere is not stationary because of the granulation and super-granulation phenomenon.[7] The flow characteristics are schematically indicated in Fig.7.6. The material moves up with a radial velocity v_r, and then flows in the transverse direction (in the form of an expanding hexagon) with an average velocity v_t. Hence the order of magnitude of the resultant "equivalent velocity" can be expressed as follows:

$$v_{eq}(\theta) \sim 0.636 \left\{ 1 + \frac{r}{M_\odot} I(\theta) \right\} - v_r \cos\theta - v_t \sin\theta. \tag{7.35}$$

The orders of magnitude of v_r and v_t have been determined as 1 km/s and 0.2 km/s, respectively. Hence

[7]Beckeres, J. M. and Nelson, G. D., *Sol. Phys.*, V. 58, 1978, p. 243; Bray, R. J. and Longhhead, R. E., *The Solar Granulation*, Champan & Hall, 1967; Cloutman, L. D., *Space Sci. Rev.*, V.2, 1980, p. 23; Küveler, G., *Sol. Phys.*, V.88, 1983, p. 13.

Table 7.2: Redshift of photons originating from the Sun at different latitudes.

θ	v_{eq} (km/s)	v_{eq} (km/s) treating the solar mass at the centre
0	0.166	0.272
$\pi/50$	0.155	0.261
$\pi/10$	0.145	0.239
$\pi/8$	0.153	0.241
$\pi/6$	0.178	0.253
$\pi/5$	0.209	0.272
$\pi/4$	0.272	0.317
$\pi/3$	0.420	0.440
$\pi/2$	0.836	0.860

$$v_{eq}(\theta) \sim 0.636 \left\{1 + \frac{r}{M_\odot} I(\theta)\right\} - \cos\theta - 0.2 \sin\theta. \tag{7.36}$$

It is not possible to determine $I(\theta)$ analytically, and the solution was found using a computer program. The following values have been used to determine $v_{eq}(\theta)$:

$$\begin{aligned} r &= 7 \times 10^8 \text{ m} \\ M_\odot &= 2 \times 10^{30} \text{ kg} \end{aligned}$$

The values of $v_{eq}(\theta)$ for different values of θ are shown in Table 7.2. Values of v_{eq} computed by treating the solar mass as concentrated at the centre are also shown for the purpose of comparison.

It is found that for larger values of θ the agreement is quite good. The resultant nature of variation of v_{eq} with $\cos\theta$ is shown in Fig.7.7, along with observations by earlier researchers.[8] The figure also shows the normalized equivalent velocity (given by subtracting the redshift at the centre). It is clear from the figure that the agreement with a large number of observational results is quite remarkable. The main outstanding problem is to explain the excess redshift at the limb. Considering the variation of v_{eq} to be only due to granulation effect the result should have appeared to be that shown in Fig.7.8. The excess redshift is explained by the velocity-dependent inertial induction.

[8] Bertolli, B. *et al.* in: *Gravitation: An Introduction to Current Research* (ed.) Louis Witten, John Wiley, 1962.

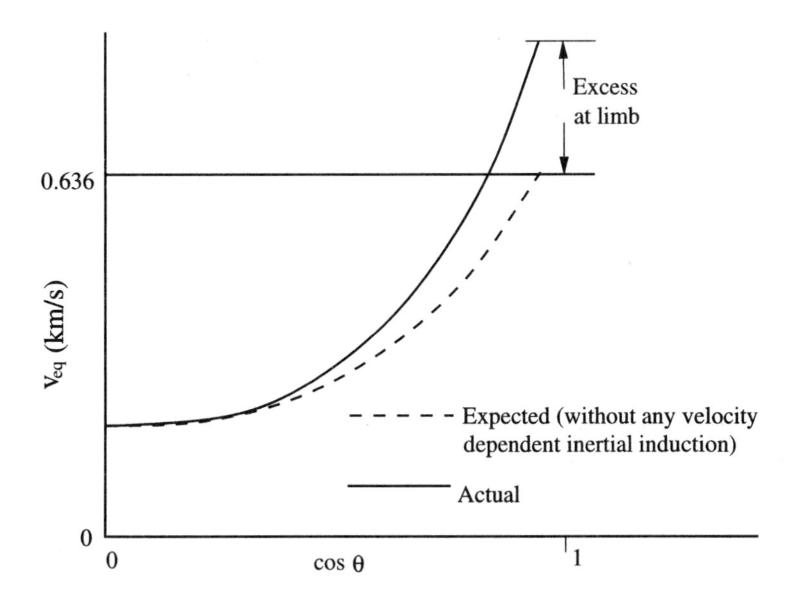

Figure 7.8: Variation in the redshift with and without the inertial induction effect.

7.5 Redshift of Photons Grazing a Massive Object

According to conventional theory, when a photon grazes a massive body, no resultant redshift is expected. While the photon approaches the object its energy increases due to the gravitational pull. This causes the photon to be blueshifted. Once the photon crosses the close approach position, it begins losing energy because the gravitational pull resists the motion. This causes the photon to be redshifted (Fig.7.9). The blueshift during the approach gets cancelled by the same amount of redshift during the recession. On the whole, the wavelength of the photon remains unaffected.

Now, if we introduce the concept of velocity-dependent inertial induction into the picture, the situation changes. Figure 7.10 shows the force on the photon due to velocity-dependent inertial induction along with the Newtonian gravitational pull. Unlike the gravitational pull, which helps the motion during the approach and then resists the motion during recession, a component of the velocity-dependent inertial induction resists the motion throughout. Therefore, a resultant redshift of the photon is expected. To simplify the analysis we may consider the matter of the massive body to be concentrated at the centre 0. At any instant when the photon is at a distance s from the centre, the magnitude of the velocity-dependent inertial drag is

$$F = \frac{G_0 M m_p}{c^2 s^2} c^2 \cos\eta |\cos\eta|,$$

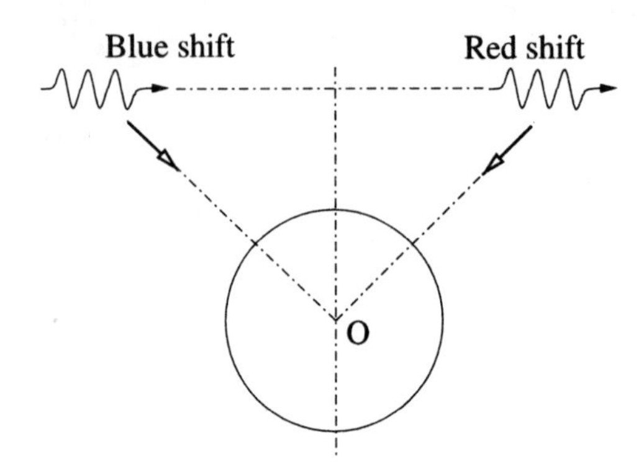

Figure 7.9: Change in the wavelength of a photon grazing past a body according to conventional physics.

where M is the mass of the object, m_p is the mass of the photon and η is the angle made by the direction of the motion with the instantaneous position line AO. The component of F in the x-direction is

$$F_x = \frac{G_0 M m_p}{s^2} \cos^2 \eta |\cos \eta|.$$

The change in energy of the photon due to the inertial induction effect when it moves a distance dx is then

$$dE = -F_x dx = -\frac{G_0 M m_p}{s^2} \cos^2 \eta |\cos \eta| dx.$$

Since $E = h\nu$, we have

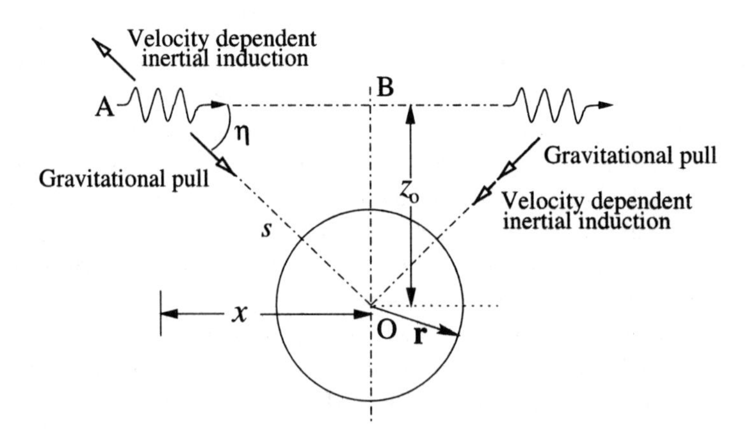

Figure 7.10: Inertial induction effect on a photon grazing past a body.

$$hd\nu = -\frac{G_0 M h\nu}{s^2 c^2} \cos^2 \eta |\cos \eta| dx,$$

or,

$$\frac{d\nu}{\nu} = -\frac{G_0 M}{c^2} \frac{\cos^2 \eta |\cos \eta|}{s^2} dx. \tag{7.37}$$

Considering the symmetry, the total change in energy of the photon grazing past the object (and, therefore, the total change in its frequency also) will be double of that for approach of the photon up to the closest position B. Thus

$$\int_{\nu}^{\nu+\Delta\nu} \frac{d\nu}{\nu} = -\frac{2G_0 M}{c^2} \int_{-\infty}^{0} \frac{\cos^3 \eta dx}{s^2}, \tag{7.38}$$

since η varies from 0 to $\pi/2$ in the range of integration and $|\cos \eta| = \cos \eta$. From Figure 7.10 we get

$$x = z_0 \cot \eta$$

and

$$s^2 = z_0 \text{cosec}^2 \eta$$

Differentiating the first of the above two relations, we obtain

$$dx = -z_0 \text{cosec}^2 \eta d\eta.$$

Substituting dx and s^2 in (7.38) and changing the limits, we have

$$\begin{aligned} \ln\left(\frac{\nu + \Delta\nu}{\nu}\right) &= +\frac{2G_0 M}{c^2} \int_{-\infty}^{0} \frac{\cos^3 \eta d\eta}{z_0} \\ &= -\frac{4G_0 M}{3c^2 z_0}. \end{aligned}$$

Finally,

$$\frac{\Delta\lambda}{\lambda} = -\frac{\Delta\nu}{\nu} = \exp\left(\frac{4G_0 M}{3c^2 z_0}\right) - 1. \tag{7.39}$$

In the case when the photon just grazes the object of radius r, the resultant fractional redshift is given by the following relation:

$$z = \frac{\Delta\lambda}{\lambda} \approx \exp\left(\frac{4G_0 M}{3c^2 r}\right) - 1. \tag{7.40}$$

Table 7.3: Redshift of photons grazing past different heavenly bodies.

Type of object	M	r	z
Jupiter	$10^{-3}M_\odot$	$r_\odot/10$	$\sim 2.7\times10^{-8}$
Sun	$M_\odot$	$r_\odot$	$\sim 2.7\times10^{-6}$
Typical white dwarf	$M_\odot$	$r_\odot/80$	$\sim 2.16\times10^{-4}$
Typical neutron star	$2M_\odot$	10 km	~ 0.5
Black hole	—	Schwarzschild radius	~ 1

Where $M_\odot$ is the solar mass and $r_\odot$ is the solar radius.

The approximate sign is used because we have assumed the whole mass of the object to be concentrated at the centre. The exact magnitude cannot be found in such a closed form of analytical solution. However, the above relation provides some idea about the order of magnitude of the redshift. The orders of magnitude for different classes of objects are given in Table 7.3.

Evidently, the redshift caused by Jupiter is too small to be detected by currently available measuring techniques. However, it is hoped that once the method of laser heterodyne spectroscopy is perfected, it will be possible to detect redshifts of the order

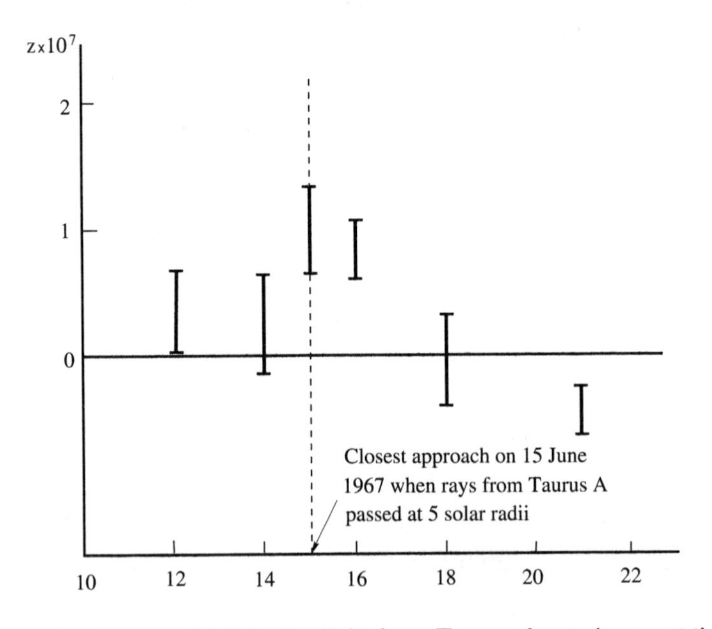

Figure 7.11: Excess redshift in the light from Taurus A grazing past the Sun.

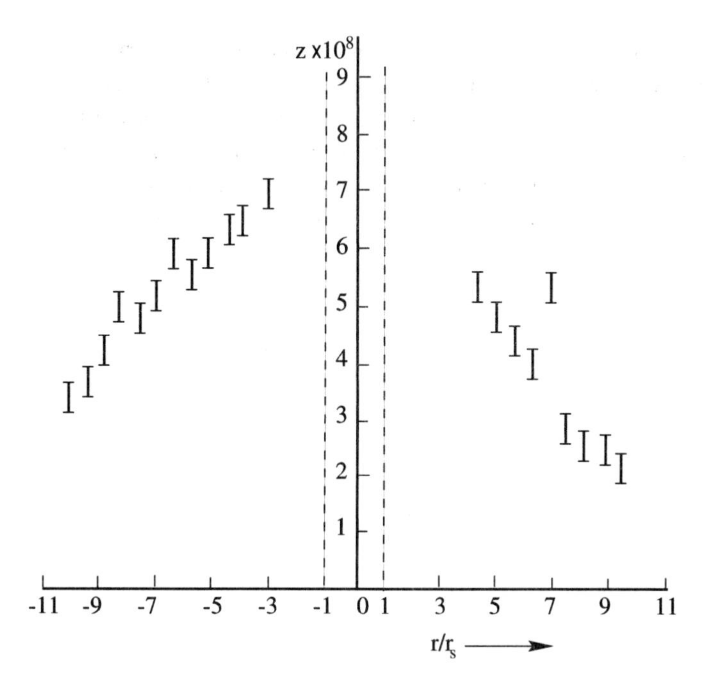

Figure 7.12: Excess redshift in the signal from Pioneer-6 grazing past the Sun.

of 10^{-8}. Then redshift during occultation by Jupiter may be detected. The redshift which is expected to be produced by velocity-dependent inertial induction due to the Sun is measurable. It is very interesting to note that reports of such unexplained redshifts exist in the literature. The first such report was made by Sadeh,[9] who reported that the 21 cm signal from Taurus A at a near occultation position by the Sun suffered a resultant redshift of 150 Hz while grazing the Sun with the nearest distance being 5 solar radii. Figure 7.11 shows the results, which are in reasonable agreement with the values estimated using (7.40) so far as the order of magnitude is concerned. Another report of unexplained redshift of signals grazing past the solar disc was made by Merat *et al.*[10]. They reported that the 2292 MHz signal from Pioneer-6 was also found to be subjected to an unexplained redshift when it passed behind the Sun. Figure 7.12 shows the variation of the redshift with the distance of signal path from the Sun's centre. The results are again in reasonable agreement with the predicted values.

There have been attempts to explain this redshift as the outcome of an interaction of the photons with the solar corona. However, the explanation in terms of the proposed existence of velocity-dependent inertial induction is most appropriate. If in future the predicted redshift of photons grazing past the Jupiter is also observed, the conventional

[9]Sadeh *et al.*, *Science*, V.159, 1968, p. 307.

[10]Merat *et al.*, *Astronomy and Astrophysics*, V.174,1974, p. 168.

explanation will fail, as there is no Jovian corona. More accurate determination of the redshift of signals grazing past the Sun should be attempted using the currently available technology, and a better matching with the prediction should be attempted. A critical analysis of the light coming from eclipsing binaries involving a white dwarf may also provide good observational data. Analysis of binary pulsar signals may also be useful in investigating the possibility of velocity-dependent inertial induction.

Chapter 8

Interaction of Matter with Matter

8.1 Introduction

VELOCITY-DEPENDENT inertial induction of a local nature between matter and matter can give rise to a number of interesting phenomena. Certain effects may not even be conceivable according to conventional physics. If the predicted effects can be detected (and if there is quantitative agreement between the observational results and theoretical predictions), the proposed theory will receive strong support. Considering the extremely weak nature of the interaction and the extremely small magnitudes of the effects, terrestrial laboratory experiments may not be possible at present. However, it is possible to investigate the motions of the solar system members and detect the presence of the predicted effects.

8.2 Inertial Induction in Some Ideal Configurations

Before taking up any real situation it is desirable to analyse some ideal cases. This will provide some insight into the new possibilities which are not present in conventional physics.

8.2.1 Force on a particle due to a rotating ring

Consider a uniform thin ring of radius r and mass per unit length σ rotating about its centre (treated as fixed in space) with an angular speed ω. We are to find out the force

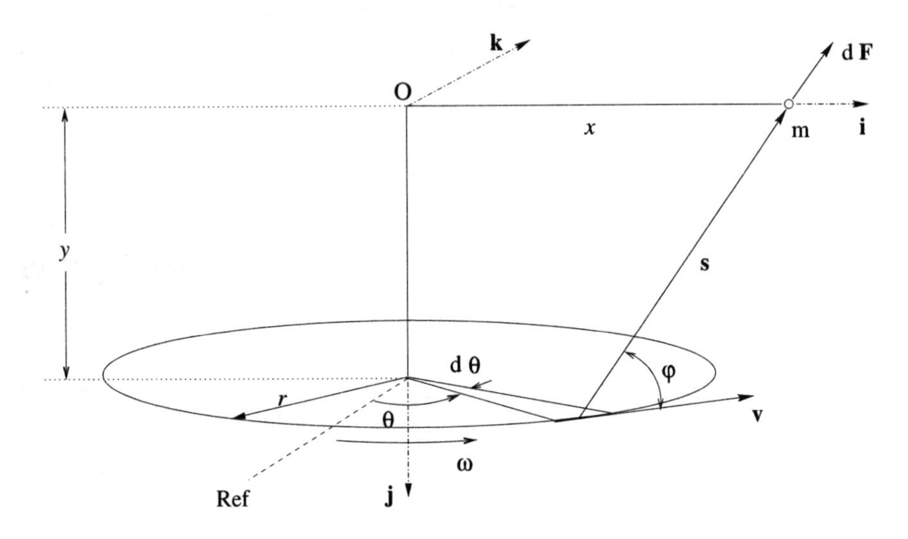

Figure 8.1: Force on a particle due to a rotating ring element.

on a particle P of mass m at a distance x from the origin on the $\hat{\mathbf{i}}$ axis. Figure 8.1 shows the configuration and the co-ordinate system used. Thus,

$$\omega = -\hat{\mathbf{j}}\omega, \tag{8.1}$$
$$\mathbf{s} = (x - r\sin\theta)\hat{\mathbf{i}} - y\hat{\mathbf{j}} + r\cos\theta\hat{\mathbf{k}}, \tag{8.2}$$
$$\mathbf{v} = \omega r(\cos\theta\hat{\mathbf{i}} + \sin\theta\hat{\mathbf{k}}) \tag{8.3}$$

where $\mathbf{v}$ is the velocity of an element of the ring at an instantaneous angle θ from the reference line, and s is the instantaneous position of the particle P. Since the mass per unit length of the ring is σ, the instantaneous force on P due to velocity dependent inertial induction from the element of the ring is given by (taking $x > r$)

$$d\mathbf{F} = \frac{Gm\sigma r d\theta}{c^2 s^2}\omega^2 r^2 \cos^2\phi.\hat{\mathbf{s}}, \tag{8.4}$$

where x is the x-co-ordinate of the particle P. From (8.2),

$$\begin{aligned} s &= \left\{(x - r\sin\theta)^2 + y^2 + r^2\cos^2\theta\right\}^{1/2} \\ &= (x^2 - 2xr\sin\theta + y^2 + r^2)^{1/2}. \end{aligned} \tag{8.5}$$

Hence the unit vector along s

$$\begin{aligned} \hat{\mathbf{s}} = & \frac{x - r\sin\theta}{(x^2 + y^2 + r^2 - 2xr\sin\theta)^{1/2}}\hat{\mathbf{i}} - \frac{y}{(x^2 + y^2 + r^2 - 2xr\sin\theta)^{1/2}}\hat{\mathbf{j}} \\ & + \frac{r\cos\theta}{(x^2 + y^2 + r^2 - 2xr\sin\theta)^{1/2}}\hat{\mathbf{k}}. \end{aligned} \tag{8.6}$$

From symmetry considerations it is obvious that the resultant force on P due to the whole rotating ring will be only in the $\hat{\mathbf{k}}$ direction (*i.e.*, the z-direction). The z-component of $d\mathbf{F}$ is

$$dF_z = \frac{Gm\sigma r d\theta}{c^2 s^2}\omega^2 r^2 \cos^2\phi \frac{r\cos\theta}{(x^2+y^2+r^2-2xr\sin\theta)^{1/2}}, \tag{8.7}$$

where

$$\begin{aligned}\cos\phi &= \frac{\mathbf{v}.\mathbf{s}}{vs} = \frac{\omega r\cos\theta(x-r\sin\theta)+\omega r^2\cos\theta\sin\theta}{\omega r(x^2+y^2+r^2-2xr\sin\theta)^{1/2}} \\ &= \frac{x\cos\theta}{(x^2+y^2+r^2-2xr\sin\theta)^{1/2}}.\end{aligned} \tag{8.8}$$

Using (8.8) in (8.7) and simplifying,

$$dF_z = \frac{Gm\sigma d\theta}{c^2}\frac{\omega^2 r^3 x^2\cos^3\theta}{(x^2+y^2+r^2-2xr\sin\theta)^{5/2}}. \tag{8.9}$$

Integrating over the whole ring (or integrating over the range $-\pi/2 < \theta < \pi/2$ and multiplying the result by 2)

$$F_z = \frac{2Gm\sigma\omega^2 r^3 x^2}{c^2}\int_{-\pi/2}^{\pi/2}\frac{\cos^3\theta d\theta}{(x^2+y^2+r^2-2xr\sin\theta)^{5/2}}. \tag{8.10}$$

After integration, the final result is as follows:

$$\begin{aligned}F_z &= \frac{2Gm\sigma\omega^2 r^3 x^2}{c^2}\left[\frac{\pi}{b^2}+2\frac{(2b^2-a^2)}{b(a^2-b^2)}\right. \\ &\left.\left\{\frac{2a}{b\sqrt{a^2-b^2}}\tan^{-1}\sqrt{\frac{a-b}{a+b}}-\frac{1}{a}\right\}\right].\end{aligned} \tag{8.11}$$

where $a = x^2+y^2+r^2$ and $b = -2xr$. Thus we find that a rotating ring will induce a transverse force on a particle (over and above the conventional gravitational pull, of course), which is not obtainable in conventional physics. This basic fact will lead to a mechanism for transferring angular momentum, as will be shown later.

8.2.2 Force on a particle due to a rotating spherical shell

The result obtained for a rotating ring can be extended to obtain the force on a particle due to the velocity-dependent inertial induction of a rotating spherical shell. Figure 8.2 shows the configuration. If σ is the mass of the spherical shell per unit surface

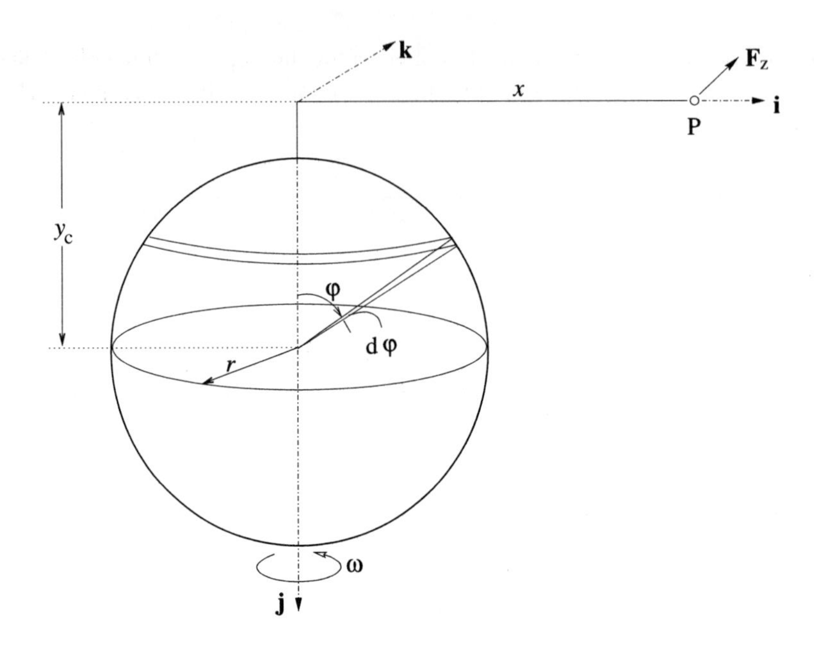

Figure 8.2: Force on a particle due to a rotating sphere.

area, then the force on the particle due to an elemental ring of the spherical shell can be expressed as follows, using (8.11):

$$
\begin{aligned}
dF_z &= \frac{2G\sigma r d\phi . m\omega^2 r^3 \sin^3\phi x^2}{c^2}\left[\frac{\pi}{b^2}+\frac{2(2b^2-a^2)}{b(a^2-b^2)}\right. \\
&\left.\left\{\frac{2a}{b\sqrt{a^2-b^2}}\tan^{-1}\sqrt{\frac{a-b}{a+b}}-\frac{1}{a}\right\}\right],
\end{aligned}
$$

where

$$a = x^2 + (y_c - r\cos\phi)^2 + r^2\sin^2\phi = x^2 + y_c^2 + r^2 - 2ry_c\cos\phi,$$

and

$$b = -2xr\sin\phi.$$

Integrating over the whole sphere, the total force on P is

$$F_z = \frac{2G\sigma r^4 m\omega^2 x^2}{c^2}\int_0^{\pi}\sin^3\phi\left[\frac{\pi}{4x^2r^2\sin^2\phi}\right.$$

$$+\frac{8x^2r^2\sin^2\phi-(x^2+y_c^2+r^2-2ry_c\cos\phi)^2}{xr\sin\phi\left\{(x^2+y_c^2+r^2-2ry_c\cos\phi)^2-4x^2r^2\sin^2\phi\right\}}$$
$$\times\left\{\frac{x^2+y_c^2+r^2-2ry_c\cos\phi}{xr\sin\phi\sqrt{(x^2+y_c^2+r^2-2ry_c\cos\phi)^2-4x^2r^2\sin^2\phi}}\right.$$
$$\tan^{-1}\left(\frac{x^2+y_c^2+r^2-2ry_c\cos\phi+2xr\sin\phi}{x^2+y_c^2+r^2-2ry_c\cos\phi-2xr\sin\phi}\right)^{1/2}$$
$$\left.\left.+\frac{1}{x^2+y_c^2+r^2-2ry_c\cos\phi}\right\}\right]d\phi \quad (8.12)$$

8.2.3 Force on a particle due to a rotating sphere

If the above result is used, the resultant force on a particle due to a rotating sphere of radius R becomes

$$F_z=\frac{2Gm\omega^2x^2}{c^2}\int_0^R\int_0^\pi\sigma(r)r^4\sin^3\phi[\;]d\phi dr \quad (8.13)$$

where $\sigma(r)$ is the density per unit volume at a distance r from the centre. The term [ ] is same as that within the square bracket in the R.H.S. of (8.12). A special case of importance is when the particle lies in the equatorial plane *i.e.*, $y_c=0$. In this situation (8.13) becomes

$$F_z=\frac{2Gm\omega^2x^2}{c^2}\int_0^R\int_0^\pi\sigma(r)r^4\sin^3\phi\left[\frac{\pi}{4x^2r^2\sin^2\phi}\right.$$
$$+\frac{8x^2r^2\sin^2\phi-(x^2+r^2)^2}{xr\sin\phi\left\{(x^2+r^2)^2-4x^2r^2\sin^2\phi\right\}}$$
$$\times\left\{\frac{x^2+r^2}{xr\sin\phi\sqrt{(x^2+r^2)^2-4x^2r^2\sin^2\phi}}\right.$$
$$\left.\left.\tan^{-1}\left(\frac{x^2+r^2+2xr\sin\phi}{x^2+r^2-2xr\sin\phi}\right)^{1/2}+\frac{1}{x^2+r^2}\right\}\right]d\phi dr \quad (8.14)$$

Figure 8.3 shows the velocity-dependent inertial induction between a rotating sphere and a particle P in the equatorial plane. ω_{rel} is the relative angular speed with respect to the particle P. It is seen that two symmetrically placed elements of the sphere produce a resultant force $d\mathbf{F}$ on P. On the other hand, the sphere is acted upon by a torque $d\tau$ due to particle P as indicated. Thus, the angular momentum of the spinning

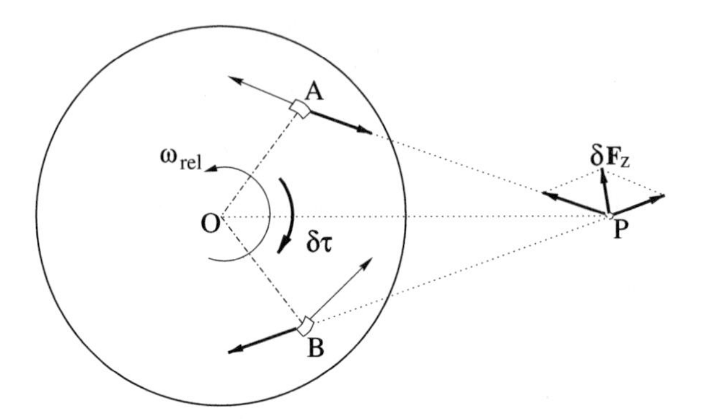

Figure 8.3: Force due to inertial induction on a particle in the equatorial plane of a rotating sphere.

sphere decreases, and the lost amount is transferred to the orbital angular momentum of the particle P. Consequently, if the rotating sphere is a planet and the particle P a satellite, there will be a constant transfer of angular momentum as a result of which P will be pushed gradually outwards (a consequence of increased angular momentum of the satellite as per Kepler's law). It is thus interesting to note that velocity-dependent inertial induction provides a mechanism for transfer of angular momentum without physical contact. No similar mechanism exists in conventional physics.

8.2.4 Torque on a rotating sphere in the vicinity of a large massive body

If there is a spinning body, A, near a large massive object B as indicated in Fig.8.4, it is easy to notice that due to the velocity of the particles of the spinning object A every particle is subjected to a force of inertial induction from body B. All these forces will result in a resisting torque on A, ultimately causing a deceleration of the spin motion. This is, again, an effect which is not considered in the domain of conventional physics. In many heavenly bodies such an effect may exist, which can be detected. Since the magnitude of the inertial induction effect is expected to be extremely small, detection

Figure 8.4: Torque on a rotating sphere near a massive object.

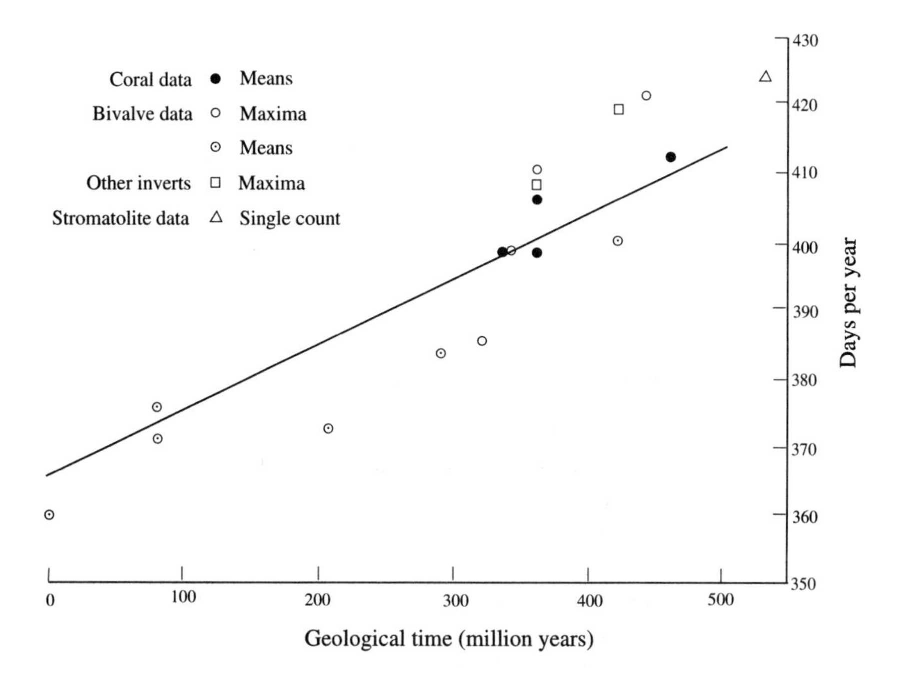

Figure 8.5: Variation of the number of days in a year.

of the effect needs observational data for a pretty long time. At present we can hope to detect it only for the spin motion of the Earth.

8.3 Secular Retardation of the Earth's Rotation

From the records of historical solar eclipses recorded in Antiquity, it was suspected that the speed of the Earth's spin had a secular retardation. Subsequently, the evidence of such a gradual decrease in the number of days in a year was found through palaeontological studies. In a number of marine creatures, a layer or material is deposited on the shell every day. The thickness of the layer also varies with the length of the day. Thus, a careful study of the cross-section of such shells can provide an estimate of the number of days in a year. Figure 8.5 shows this variation as obtained from palaeontological studies.

Considerable work has been done on the subject, and the publications are too many to be exhaustively listed here.[1] It should be mentioned at this stage that there are other

---

[1]Munk, W. H. and McDonald, G. J. F., *The Rotation of the Earth*, Cambridge Univ. Press, 1960; Dicke, R. H., in: *The Earth-Moon System*, (eds.) B. G. Marsden and A. G. W. Cameron, Plenum, New York, 1966;
Rosenberg, G. D. and Runcorn, S. K., *Growth Rhythms and the History of the Earth's Rotation*, John Wily,

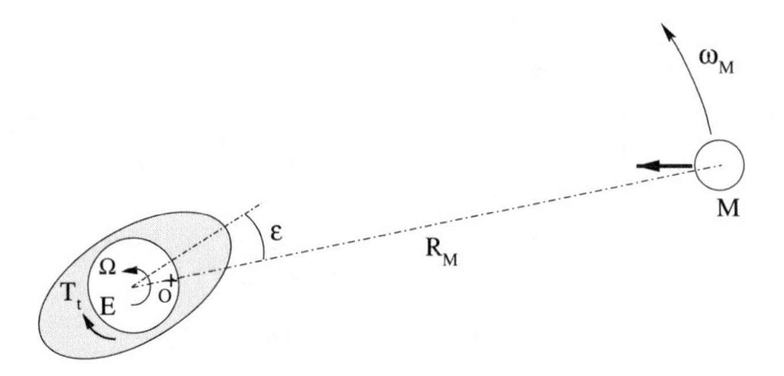

Figure 8.6: Effect of the tidal bulge.

types of non-secular fluctuations which cause the length of a day to vary. However, we are considering here the long-term true secular deceleration of the Earth's spin. Because of the many other perturbing factors, it is difficult to estimate the secular change from observations taken over a few decades only. According to the conventional theory the only major contributing factor is tidal friction. Figure 8.6 shows the Earth-Moon system and the tidal bulges (of course, greatly exaggerated). The angular velocity of the orbital motion of the Moon around the Earth, ω_M, is much smaller than the angular velocity of the Earth's spin, Ω. Ideally the tidal bulge should have been aligned with the line EM joining the centres of the Earth and the Moon. So the speed with which the bulge rotates around the Earth is equal to ω_M. The relative motion of the Earth with respect to this bulge generates a frictional torque, T, on the Earth resisting the relative spin (Fig.8.6). This torque is considered to be the source of secular retardation. If I is the moment of inertia of the Earth, then

$$\dot{\Omega} = -\frac{T_t}{I}. \tag{8.15}$$

Because of the angular shift ϵ of the bulge as shown in the figure, the gravitational pull on the Moon is also slightly tilted away from the line EM as shown. This is so because the nearer half of the bulge exerts a greater pull with a component in the forward direction (*i.e.*, in the direction of ω_M). The component of the pull due to the other half of the bulge in the backward direction is smaller because of the greater distance of the bulge from the Moon. The resultant forward component on the Moon causes its angular momentum (of orbital motion around the Earth) to increase. The loss of angular momentum of the Earth is compensated by the gain in the angular momentum of the orbital motion of the Moon. Thus, the total angular momentum of the Earth-Moon system is conserved. The Earth-Moon system rotates about the

---

London, 1975;
Stacey, F. D., *Physics of the Earth*, John Wily, New York, 1977;
Melchior, P., *The Tides of the Planet Earth*, Pergamon Press, London, 1978;
McElhinny, M. W., *The Earth : Its Origin, Structure and Evolution*, Academic Press, London, 1979.

common centre of gravity O (Fig.8.6). The Earth being much more massive than the Moon, O is very near to the Earth's centre E, as indicated. For rough calculations, we may use $r_M \approx OM$. We also assume the orbit to be circular. Now the orbital velocity and distance of the Moon are such that the centripetal acceleration is equal to the acceleration generated by the gravitational acceleration of the Earth. (In this approximate treatment we neglect the inclination of this pull, as it is extremely small). Thus

$$m_M \omega_M^2 R_M \approx \frac{G_0 m_E m_M}{R_M^2},$$

or,

$$\omega_M = \left(\frac{G_0 m_E}{R_M^3}\right)^{1/2}. \tag{8.16}$$

Now, the orbital angular momentum of the Moon can be written as

$$L_M \approx m_M \omega_M R_M^2.$$

Using (8.16)

$$L_M \approx m_M \left(G_0 m_E\right)^{1/2} R_M^{1/2}. \tag{8.17}$$

The rate at which the Earth loses its angular momentum must be equal to the rate at which the orbital angular momentum of the Moon increases. Hence

$$-I\dot{\Omega} = \frac{dL_M}{dt} = \frac{1}{2} m_M \left(\frac{G m_E}{R_M}\right)^{1/2} \dot{R}_M. \tag{8.18}$$

Again using (8.16) the angular momentum can be expressed as

$$L_M \approx m_M \left(G_0 m_E\right)^{2/3} \omega_M^{-1/3},$$

and following the same reasoning

$$-I\dot{\Omega} \approx -\frac{1}{3} m_M \left(\frac{G m_E}{\omega_M^2}\right)^{2/3} \dot{\omega}_M. \tag{8.19}$$

Hence, $\dot{\Omega}$, $\dot{\omega}_M$ and $\dot{R}_M$ are related, and if any one of them is determined from observations, the other two can be calculated using (8.18) and (8.19).

As already mentioned, determination of the secular changes in Ω, ω_M and R_M is possible only when data over a long period of time are available. Even very accurate data over a relatively short period of only a few decades may not indicate the secular change because of the presence of predominating fluctuations of shorter periods. So, $\dot{\Omega}$ can only be estimated with reasonable degree of accuracy and $\dot{\omega}_M$ and $\dot{R}_M$ can be determined. The currently accepted values of the secular changes are as follows:

$$\begin{aligned}\dot{\Omega} &\sim -6 \times 10^{-22} \text{ rad s}^{-2},\\ \dot{\omega_M} &\sim -1.3 \times 10^{-23} \text{ rad s}^{-2},\\ \dot{R_M} &\sim -1.3 \times 10^{-9} \text{ m s}^{-2}.\end{aligned}$$

There is a small effect due to the tidal phenomenon caused by the Sun. However, it can be ignored for a simple analysis without introducing much error.

Though at a first glance the above explanation of the phenomenon appears plausible, serious problems emerge on closer scrutiny. It is seen that the tidal friction phenomenon causes the Moon to recede. It can be also shown that tidal friction is

$$T_t \propto R_M^{-6}.$$

The amplitude of the tide can also be shown to be proportional to R_M^{-3}. Therefore the effects of both the magnitude of the tidal friction torque and the amplitude of tides were much stronger in the past when R_M was smaller than at present. However, even if we assume the rate $\dot{R}_M$ to be constant at the current value, then 1300 million years ago, the Earth-Moon distance should have been

$$(R_M - 1300 \times 10^6 \times 365 \times 24 \times 3600 \times 1.3 \times 10^{-9})\text{m},$$

where $R_M = 384.4 \times 10^6$ m, the present distance to the Moon. Thus, 1300 million years ago R_M was only 331.1×10^6 m and this close approach would have been devastating for both the Earth and the Moon. This does not agree with the geological evidence. Sedimentary data indicates the presence of the tidal phenomenon during the last 3500 million years; but there is no evidence of either very strong tidal amplitude or any catastrophic event. Explanations have been attempted by suggesting that the land mass configurations of the continents were such that the impact of the tidal friction was much less. However, it has not been possible to substantiate this by any dependable analysis. Without going into further details, it can be said with confidence that the conventional theory is far from satisfactory, and that it faces insurmountable difficulties. This is reflected in a comment by two noted experts in the subject, Calame and Mulholland:[2] "We must abandon the habit of treating the unexplained acceleration as being entirely of tidal origin and search for other causes that might contribute to it."

8.4 Explanation from Velocity Dependent Inertial Induction

In Section 8.2.4 it has already been shown how a spinning body can slow down in the vicinity of a massive object according to the proposed theory of velocity-dependent

[2]Calame, O. and Mulholland, J. D., in: *Tidal Friction and the Earth's Rotation*, (eds.) P. Brosche and J. Sundermann, Springer Verlag, Berlin, 1978.

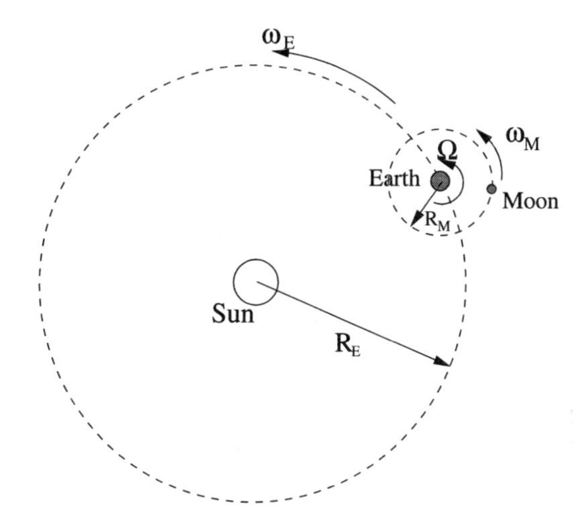

Figure 8.7: The Sun-Earth-Moon system.

inertial induction. Therefore, we should investigate if a part of the secular retardation of the Earth's spin and of the Moon's orbital motion can be due to inertial induction by the Sun. If the Sun's influence can produce a braking torque on the spinning Earth, a similar effect should be present in the case of the spinning Mars also.

Figure 8.7 shows a simplified model of the Sun-Earth-Moon system. The effect of the tilt of the Earth's axis will be considered in the analysis. Calculation shows that the effects due to other planets in the solar system are much smaller than those due to the Sun and the Moon and, therefore, can be neglected. Moreover, since most of the Sun's mass is concentrated near/around its centre (see Fig.7.5) the effect of Sun's spin has not been considered in this analysis. The various effects due to velocity-dependent inertial induction in the Sun-Earth-Moon system are as follows:

1. A braking torque on the spinning Earth due to the Sun. This results in a retardation of the Earth's spin, Ω, and a decrease in the Earth's orbital speed ω_E associated with an increase in R_E, Earth's orbital radius. The order of magnitude of $\dot{\omega}_E$ and $\dot{R}_E$ are, however, very small compared to other secular acceleration and retardation terms, and can be neglected in an approximate analysis.

2. A resisting torque on the Earth-Moon system (orbiting about common C.G) due to the Sun. This results in a loss in the angular momentum causing an acceleration in the Moon's orbital speed ω_M. This effect also makes relatively very small contributions to $\dot{\omega}_M$ and $\dot{R}_E$ which are, in any case, neglected.

3. A braking torque on the spinning Earth due to velocity-dependent inertial induction of the Moon. This results in a retardation of the Earth's spin and a retardation of

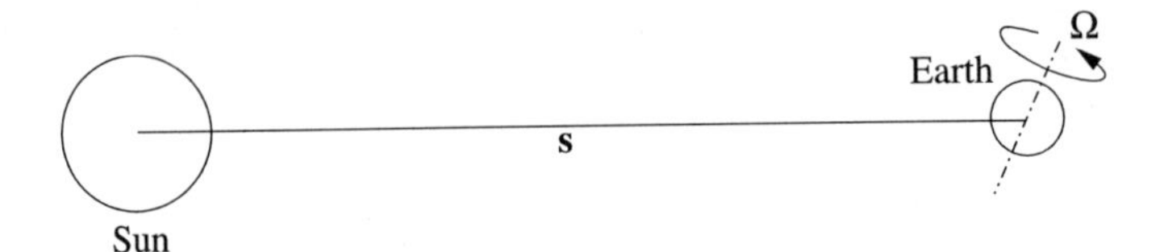

Figure 8.8: The Sun and the Earth with tilted axis.

the Moon's orbital motion, ω_M. The magnitude of this effect is also very small and can be ignored in this approximate analysis.

The effect of the tidal friction is superimposed on all the above-mentioned velocity-dependent inertial induction effects among the three interacting bodies. The magnitudes of the above-mentioned effects will now be determined; we will start with the inertial induction effect of the Sun on the spinning Earth.

8.4.1 Secular retardation of Earth's spin due to velocity dependent inertial induction of the Sun

The Earth's orbit is assumed to be approximately circular and the configuration is as shown in Fig.8.8. The lines joining points on the Earth to the Sun are all assumed to be parallel to s, the position vector of the Earth with respect to the Sun. As already mentioned, s is constant and equal to R_E. Let us now consider an elemental mass of the Earth dm_E as indicated in Fig.8.9. Using a spherical co-ordinate system, the position co-ordinates of dm_E are r, θ and ϕ as shown.
The velocity of the element is

$$\mathbf{v} = \mathbf{\Omega} \times \mathbf{r},$$

where $\mathbf{\Omega} = \Omega\hat{\mathbf{k}}$ is the angular velocity of the Earth and $\mathbf{r} = r\sin\theta\cos\phi\hat{\mathbf{i}}+r\sin\theta\sin\phi\hat{\mathbf{j}}+r\cos\theta\hat{\mathbf{k}}$ is the position vector of the element. Thus

$$\mathbf{v} = -\Omega r(\sin\theta\sin\phi\hat{\mathbf{i}} - \sin\theta\cos\phi\hat{\mathbf{j}}). \tag{8.20}$$

From the figure we can also express the unit vector in the direction of $\mathbf{v}$ as follows:

$$\hat{\mathbf{v}} = -\sin\phi\hat{\mathbf{i}} + \cos\phi\hat{\mathbf{j}}. \tag{8.21}$$

The unit vector in the direction of s can be expressed as

$$\mathbf{s} = \sin\psi\cos\delta.\hat{\mathbf{i}} + \sin\psi\sin\delta.\hat{\mathbf{j}} + \cos\psi.\hat{\mathbf{k}}. \tag{8.22}$$

The force on the element dm_E from the velocity-dependent inertial induction due to the solar mass is given by

$$d\mathbf{F} = -\frac{G_0 m_s dm_E}{c^2 s^2}\Omega^2 r^2 \sin^2\theta.(\hat{\mathbf{s}}.\hat{\mathbf{v}}).|\hat{\mathbf{s}}.\hat{\mathbf{v}}|\hat{\mathbf{s}}, \tag{8.23}$$

where m_s is the mass of the Sun, and

$$dm_E = \rho(r)r^2 dr \sin\theta d\theta d\phi, \tag{8.24}$$

with $\rho(r)$ as the density of the Earth at a distance r from the centre. Now the torque of this force on the spinning Earth can be expressed as:

$$\begin{aligned} dT_z &= r_x dF_y - r_y dF_x \\ &= r\sin\theta\cos\phi.dF_y - r\sin\theta.\sin\phi dF_x \\ &= Ar\sin\theta\sin\psi\sin(\delta-\phi), \end{aligned} \tag{8.25}$$

where,

$$A = -\frac{G_0 m_s \Omega^2}{c^2 R_E^2}\left\{\rho(r)r^4\sin^3\theta.(\hat{\mathbf{s}}.\hat{\mathbf{v}}).|\hat{\mathbf{s}}.\hat{\mathbf{v}}|drd\theta d\phi\right\}.$$

From (8.21) and (8.22)

$$\hat{\mathbf{s}}.\hat{\mathbf{v}} = \sin\psi\cdot\sin(\delta-\phi).$$

Using this in (8.25) and integrating over the whole Earth the total torque T_z on the spinning Earth becomes

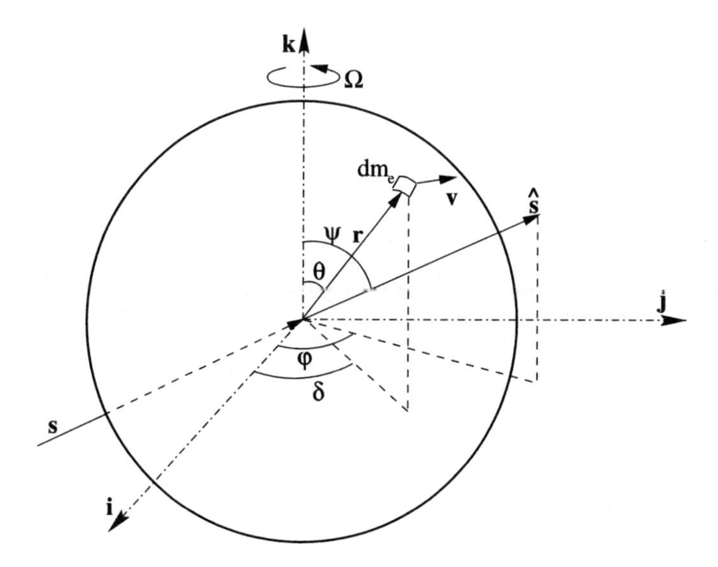

Figure 8.9: Force on an element of the Earth due to the inertial induction effect of the Sun.

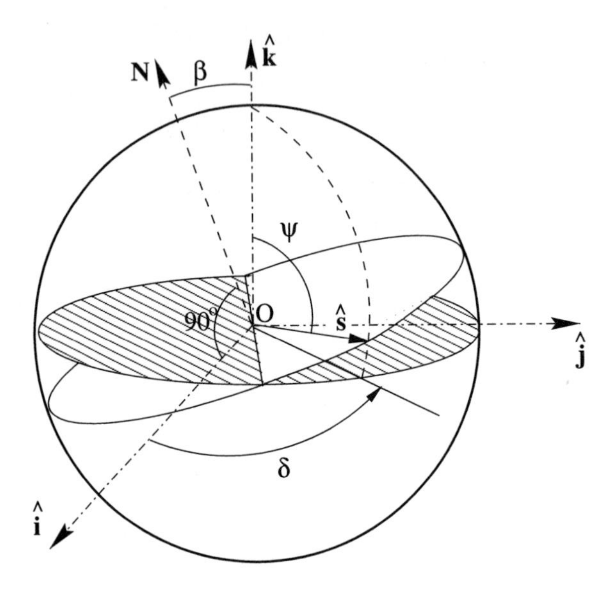

Figure 8.10: Variation in orientation of the Earth's spin axis.

$$T_z = -\frac{G_0 m_s \Omega^2}{c^2 R_E^2} \sin^3\psi \int_0^{r_E} dr \int_0^{\pi} d\theta \int_0^{2\pi} d\phi \left\{\rho(r) r^5 \sin^4\theta . \sin^2(\delta - \phi) . |\sin(\delta - \phi)|\right\}, \tag{8.26}$$

where r_E is the Earth's radius. (The negative sign is to indicate that T_z opposes the spin). Now it should be noted that as the Earth revolves round the Sun, the orientation of $\hat{s}$ with respect to the co-ordinate system changes as the z-axis is aligned with the axis of the Earth, which remains oriented to a fixed direction in space. Figure 8.10 shows the relation among the various quantities. The shaded plane is the cross-section of a sphere with the Earth's centre as the centre and unit radius cut by the plane of the ecliptic (*i.e.*, the plane in which Earth's orbit lies). Thus, at any instant, $\hat{s}$ is the unit vector and during the complete cycle of Earth's revolution around the Sun $\hat{s}$ rotates in the shaded plane (as $\hat{s}$ is in the direction of **s** which is always in the plane of the ecliptic). The normal (ON) to the shaded cross-section (*i.e.*, the plane of the ecliptic) makes a fixed angle β with the z-axis and, in case of the Earth, $\beta = 23.5^0$. Now from Fig.8.10, the following relation holds:

$$\cos\psi \cos\beta = \sin\psi \cdot \sin\delta \cdot \sin\beta \tag{8.27}$$

This is derived by first expressing the components of the unit vector $\hat{s}$ as follows:

$$\begin{aligned}\hat{s}_x &= \sin\psi\cos\delta\\ \hat{s}_y &= \sin\psi\sin\delta\\ \hat{s}_z &= \cos\psi\end{aligned}$$

Now since ŝ is perpendicular to ON, the sum of the components $\hat{s}_x$, $\hat{s}_y$ and $\hat{s}_z$ along ON must be equal to zero. Thus

$$s_x \cos 90^0 + s_y \cos(90^0 + \beta) + s_z \cos\beta = 0.$$

From the above equation we get (8.27). Hence,

$$\tan\psi = \cot\beta / \sin\delta,$$

or,

$$\sin^3\psi = \left\{\cot^2\beta / (\cot^2\beta + \sin^2\delta)\right\}^{3/2}.$$

Substituting the above relation in (8.26) we get T_z as a function of δ, as follows:

$$\begin{aligned}T_z &= -\frac{G_0 m_s \Omega^2}{c^2 R_E^2}\left\{\cot^2\beta/(\cot^2\beta+\sin^2\delta)\right\}^{3/2}\int_0^{r_E} dr \int_0^{\pi} d\theta \int_0^{2\pi} d\phi\\ &\quad \left\{\rho(r) r^5 \sin^4\theta .\sin^2(\delta-\phi).|\sin(\delta-\phi)|\right\}.\end{aligned} \tag{8.28}$$

During the Earth's orbiting motion δ varies from 0 to 2π, and the average torque resisting the Earth's spin is

$$(T_z)_{av} = \frac{1}{2\pi}\int_0^{2\pi} T_z d\delta. \tag{8.29}$$

The density of the Earth as a function of r can be expressed as follows:

$$\rho(\bar{r}) = \begin{cases} (18 - 10\bar{r}) \times 10^3 & \text{for } 0 \le \bar{r} \le 0.2\\ (13.143 - 5.714\bar{r}) \times 10^3 & \text{for } 0.2 \le \bar{r} \le 0.55\\ (9.667 - 6.667\bar{r}) \times 10^3 & \text{for } 0.55 \le \bar{r} \le 1\end{cases}$$

where $\bar{r} = r/r_E$. Using the above density function and the following data

$$\begin{aligned}G_0 &= 6.67 \times 10^{-11}\ \text{m}^3.\text{kg}^{-1}\text{s}^{-2}\\ m_s &= 1.99 \times 10^{30}\ \text{kg}\\ \Omega_s &= 7.2722 \times 10^{-5}\ \text{rad s}^{-1}\end{aligned}$$

$$\begin{aligned} r_E &= 6.378 \times 10^6 \quad \text{m} \\ R_E &= 1.496 \times 10^{11} \quad \text{m} \\ \beta &= 23.5^0 \\ c &= 3 \times 10^8 \quad \text{m s}^{-1} \end{aligned}$$

$(T_z)_{av}$ can be computed.[3] It becomes equal to

$$-4.414 \times 10^{16} \quad \text{N m}.$$

The moment of inertia of the Earth is known to be

$$I_E = 8.04 \times 10^{37} \quad \text{kg m}^2.$$

From the above two quantities the secular retardation of the Earth's spin is derived as follows:

$$\dot{\Omega} = -5.5 \times 10^{-22} \quad \text{rad s}^{-2}.$$

It is truly amazing that we get a value so close to the observational estimate of -6×10^{-22} rad s^{-2}. It is seen that only a very small part -0.5×10^{-22} rad s^{-2} is produced by the tidal friction. The value of $\dot{\Omega}$ is derived from Fig.8.5 as follows. The number of days, Ω and ω_E are related by the following equation:

$$N_y = \Omega/\omega_E - 1.$$

Since $\dot{\omega}_E$ is very small compared to $\dot{\Omega}$ we can write

$$\dot{\Omega} \approx \dot{N}_y \omega_E.$$

From Fig.8.5 the value of $\dot{N}_y$ is -3×10^{-45} s^{-1} which yields

$$\dot{\Omega} \approx -6 \times 10^{-22} \quad \text{rad s}^{-2}$$

Since only about -0.5×10^{-22} rad s^{-2} of the secular retardation is to be explained in terms of the tidal friction, the magnitude of the torque due to the tidal friction is

$$T_t \sim 0.4 \times 10^{16} \quad \text{N m}.$$

The torque on the rotating Earth-Moon system (like a dumbbell with very unequal lobes) due to the Sun is a fluctuating quantity, and the average value of the resisting torque[4] is

[3] The detailed computation is omitted here as it is a routine exercise.

[4] Ghosh, A., *Pramana (Jr. of Physics)*, V.26, No.1, 1986, p.1.

$$T_{EM} \approx -0.42 \frac{G_0 m_s m_E}{c^2 R_E^2} \left(\frac{m_M}{m_M + m_E} \right)^2 \omega_M^2 R_M^3, \tag{8.30}$$

where m_M is the mass of the Moon, R_M is the distance of the Moon from the Earth and ω_M is the orbital angular velocity of the Moon. The angular momentum of the Earth-Moon system about the common C.G. is

$$L_{EM} = \frac{m_E m_M}{m_E + m_M} R_M^2 \omega_M. \tag{8.31}$$

This is a more accurate expression compared to (8.17). Again we know

$$\frac{G_0 m_E m_M}{R_M^2} = m_M \omega_M^2 R_M . \frac{m_M}{m_E + m_R}.$$

Using the above equation R_M^2 is replaced in terms of ω_M in (8.31) to yield L_{EM} as a function of ω_M only as follows:

$$L_{EM} = \frac{m_E m_M}{(m_E + m_M)^{1/3}} G_0^{2/3} \omega_m^{-1/3}.$$

Differentiating with respect to time, we obtain

$$\begin{aligned} \dot{L}_{EM} = T_{EM} &= \frac{m_E m_M}{(m_E + m_M)^{1/3}} G_0^{2/3} \left(-\frac{1}{3} \omega_M^{-4/3} \right) \dot{\omega}_M \\ &= -\frac{L_{EM}}{3\omega_M} \dot{\omega_M}, \end{aligned}$$

or,

$$\dot{\omega}_M^s = -3\omega_M T_{EM} / L_{EM}. \tag{8.32}$$

Hence the average value of the secular acceleration of the Moon's orbital motion due to the velocity-dependent inertial induction of the Sun can be expressed approximately as

$$(\dot{\omega}_M^s)_{av} \approx 0.27 \times 19^{-23} \quad \text{rad s}^{-2}$$

However, the tidal friction will cause the angular motion of the Moon to decelerate. A tidal friction of 0.4×10^{16} Nm will produce

$$\dot{\omega}_M^t \approx -0.11 \times 10^{-23} \quad \text{rad s}^{-2},$$

and the resultant average secular acceleration of the orbital motion of the Moon will be

$$(\dot{\omega}_M)_{av} = (\dot{\omega}_M^s)_{av} + \dot{\omega}_M^t \approx 0.16 \times 10^{-23} \quad \text{rad s}^{-2}.$$

The value of $\dot{\omega}_M$ due to the inertial induction of the spinning Earth on the Moon is very small compared to the above-mentioned effects and is neglected. Again

$$\begin{aligned}\dot{R}_M &= -\frac{2}{3}\frac{R_M}{\omega_M}.\dot{\omega}_M \\ &\approx -0.15 \times 10^{-9}\text{m s}^{-1}.\end{aligned}$$

The effect of the Moon on the Earth's spin due to velocity-dependent inertial induction is only about 5% of that of the Sun, and it is neglected in estimating both $\dot{\Omega}$ and $\dot{\omega}_M$.

Using the above results, the apparent solar and lunar acceleration can be estimated as follows:

$$\dot{\omega}_s^{ap} = \dot{\omega}_E - \dot{\Omega}\frac{\omega_E}{\Omega} \sim 1.65 \times 10^{-24} \quad \text{rad s}^{-2},$$

and

$$\dot{\omega}_M^{ap} = \dot{\omega}_M - \dot{\Omega}\frac{\omega_M}{\Omega} \sim 2.3 \times 10^{-23} \quad \text{rad s}^{-2}.$$

The most significant result is that $\dot{R}_M$ is negative and the magnitude is about one tenth of the value derived using the tidal friction theory only. Hence, we find that the Moon is actually approaching the Earth with a very small speed, and there is no close-approach problem.

8.5 Secular Retardation of Mars

In conventional physics, tidal friction is considered to be the only mechanism for secular retardation of the Earth's spin. Since there is no massive satellite in the case of Mars, any such secular change in the spin motion of Mars is not expected. However, in the previous section it was shown that the major contribution to the secular retardation of the Earth's spin comes from velocity-dependent inertial induction of the Sun on the Earth. Since this is true for Mars also, one can expect a secular retardation of its spin motion. In this section a similar analysis is carried out to determine $\dot{\Omega}_{\text{Mars}}$ of Mars.

The relevant data in case of the sun-Mars system are as follows:

$$\begin{aligned}R_{Mars} &= 1.524 \quad \text{AU} \\ m_{Mars} &= 6.42 \times 10^{23} \quad \text{kg} \\ r_{Mars} &= 3.386 \times 10^{6} \quad \text{m} \\ \Omega_{Mars} &= 7.0886 \times 10^{-5} \quad \text{rad s}^{-1} \\ \beta_{Mars} &= 25.2^{o}.\end{aligned}$$

The orbit of Mars is also assumed to be circular, and the density function is as follows:[5]

$$\begin{aligned}\rho(r) &= 5800\text{kg m}^{-3} \quad \text{for} \quad 0 \le r \le 2.136 \times 10^6 \quad \text{m};\\ \rho(r) &= 3600\text{kg m}^{-3} \quad \text{for} \quad 2.136 \times 10^6 m \le r \le 3.386 \times 10^6 \quad \text{m}.\end{aligned}$$

With this density function the moment of inertia of Mars (assuming it to be approximately spherical) is

$$I_{\text{Mars}} \approx 2.85 \times 10^{36} \quad \text{kg m}^2.$$

The average torque on spinning Mars due to velocity-dependent inertial induction is estimated using (8.28), with numerical computation. We get

$$(T_z)_{av} \approx -3.56 \times 10^{14} \quad \text{Nm}.$$

The secular retardation (using the values of $(T_z)_{av}$ and I_{Mars}) becomes

$$\dot{\Omega}_{\text{Mars}} \approx -1.25 \times 10^{-22} \text{rad s}^{-2}.$$

This is a quantity smaller than that obtained for the earth, but the orders of magnitude are about the same. At the present time, there is no observational data on $\dot{\Omega}_{\text{Mars}}$. Perhaps there has been no effort to detect it, because it is not expected from conventional physics. However, if in future $\dot{\Omega}_{\text{Mars}}$ is observationally found to be around this magnitude, it will be a very strong case supporting the theory of velocity-dependent inertial induction.

8.6 Secular Acceleration of Phobos

Though there is no large satellite for the planet Mars, there are two very small natural satellites. One of these, Phobos, is quite near the planet and orbits round the Mars at a fast rate like many artificial satellites of the earth. Like most natural satellites, Phobos lies in the equatorial plane of Mars and its direction of rotation is the same as that of the planet, as indicated in Fig.8.11.

There are two possible ways in which the orbital motion of Phobos may be affected. There will be a secular acceleration of its orbital motion due to the velocity-dependent inertial induction of the sun with the rotating Mars-Phobos system. Following a similar approximate analysis as done in the case of the earth-moon system, the average secular acceleration of Phobos due to the sun can be expressed as[6]

[5]Cook, *Interiors of Planets*, Cambridge University Press, 1980, p.195.
[6]Ghosh, A., *Pramana (Jr. of Physics)*, V.26, No.1, 1986, p.1.

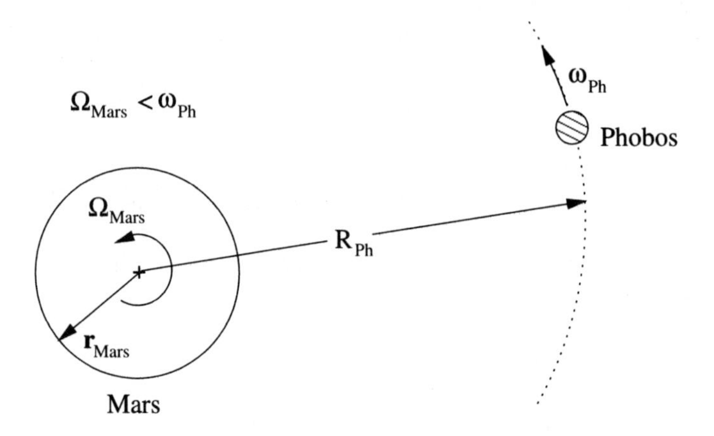

Figure 8.11: The Mars-Phobos system.

$$\left(\dot{\omega}^{s}_{ph}\right)_{av} \approx \frac{4}{\pi}\frac{G_0 m_s}{c^2 R^2_{\text{Mars}}}\left(\frac{m_{\text{Ph}}}{m_{\text{Mars}}}\right)\omega^2_{ph} R_{ph}, \tag{8.33}$$

where R_{Mars} is the radius of the Mars' orbit (assumed circular), R_{Ph} is the average distance of Phobos from the centre of Mars, ω_{Ph} is the angular velocity of Phobos around Mars, m_{Ph} is the mass of Phobos and m_{Mars} is the mass of Mars. The approximation has also used the condition the $m_{\text{Ph}} << m_{\text{Mars}}$. Inserting the values

$$\begin{aligned}
\omega_{\text{Ph}} &= 2.28 \times 10^{-4} \ \text{rad s}^{-1} \\
m_{\text{Mars}} &= 6.42 \times 10^{23} \ \text{kg} \\
R_{\text{Ph}} &= 9.133 \times 10^{6} \ \text{m} \\
m_{\text{Ph}} &= 17.4 \times 10^{15} \ \text{kg} \\
\Omega_{\text{Mars}} &= 7.088 \times 10^{-5} \ \text{rad s}^{-1}
\end{aligned}$$

in (8.33), we get the average value

$$\left(\dot{\omega}^{s}_{\text{Ph}}\right)_{av} \sim 5.7 \times 10^{-28} \text{rad s}^{-1}.$$

The other source of secular acceleration of Phobos is the velocity-dependent inertial drag on the satellite due to the planet. The force on a satellite due to its relative rotation with respect to its planet can be derived, as shown below.

Figure 8.12 shows the details of the configuration. The satellite P is orbiting around the planet in a circular orbit of radius R with an angular velocity ω. The orbit lies in the equatorial plane of the planet and the planet is spinning with an angular velocity Ω as indicated. The mass of the satellite is m_P.

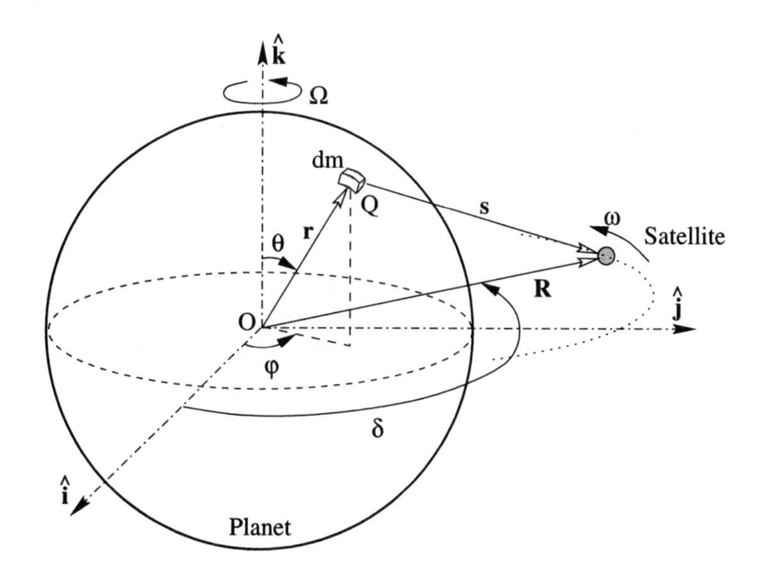

Figure 8.12: Force on the Martian satellite due to inertial induction.

Now let us consider an elemental mass dm of the planet. The force on the satellite due to velocity-dependent inertial induction from dm can be expressed as

$$d\mathbf{F} = -\frac{G_0 dm.m_P v_{rel}^2}{c^2 s^2} \cos\phi.|\cos\psi|\hat{\mathbf{s}}, \tag{8.34}$$

where $\cos\psi = \hat{\mathbf{s}}.\hat{\mathbf{v}}_{rel}$ and $\mathbf{v}_{rel}$ is the relative velocity of the satellite with respect to dm at point Q. So,

$$\mathbf{v}_{rel} = \mathbf{v}_P - \mathbf{v}_Q, \tag{8.35}$$

where $\mathbf{v}_P$ is the velocity of the satellite and $\mathbf{v}_Q$ is the velocity of the elemental mass dm. Now from Fig.8.12

$$\begin{aligned} \omega &= \hat{\mathbf{k}}\omega \\ \mathbf{\Omega} &= \hat{\mathbf{k}}\Omega \\ \mathbf{v}_Q &= \mathbf{\Omega} \times \mathbf{r} \end{aligned}$$

and

$$\mathbf{r} = \hat{\mathbf{i}} r \sin\theta\cos\phi + \hat{\mathbf{j}} r \sin\theta \sin\phi + \hat{\mathbf{k}} r \cos\theta,$$

where $\mathbf{r}$ is the position of Q with respect to the centre of the planet O. Using the above relations, we find

$$\begin{aligned} \mathbf{v}_Q &= -\hat{\mathbf{i}}(\Omega r \sin\theta \sin\phi) + \hat{\mathbf{j}}(\Omega r \sin\theta\cos\phi) \\ &= \Omega r \sin\theta(-\hat{\mathbf{i}}\sin\phi + \hat{\mathbf{j}}\cos\phi). \end{aligned} \tag{8.36}$$

Similarly

$$\mathbf{v}_P = \omega \times \mathbf{R},$$

where $\mathbf{R} = \hat{\mathbf{i}}R\cos\delta + \hat{\mathbf{j}}R\sin\delta$ is the position vector of the satellite. Thus

$$\mathbf{v}_P = \omega R(-\hat{\mathbf{i}}\sin\delta + \hat{\mathbf{j}}\cos\delta). \tag{8.37}$$

Using (8.36) and (8.37) in (8.35) we get

$$\begin{aligned} \mathbf{v}_{rel} &= (\Omega r \sin\theta\sin\phi) - \omega R\sin\delta)\hat{\mathbf{i}} \\ &\quad +(-\Omega r\sin\theta\cos\phi + \omega R\cos\delta)\hat{\mathbf{j}}. \end{aligned} \tag{8.38}$$

Again from Figure 8.12,

$$\begin{aligned} \mathbf{s} &= \mathbf{R} - \mathbf{r} \\ &= (R\cos\delta - r\sin\theta\cos\phi)\hat{\mathbf{i}} + (R\sin\delta - r\sin\theta\sin\phi)\hat{\mathbf{j}} - r\cos\theta\hat{\mathbf{k}}. \end{aligned}$$

So,

$$s^2 = R^2 + r^2 - 2rR\sin\theta\; cos(\phi - \delta). \tag{8.39}$$

Now

$$\cos\phi = \frac{\mathbf{s}\cdot\mathbf{v}_{rel}}{s v_{rel}}.$$

Substituting **s**, $\mathbf{v_{rel}}$, s and v_{rel} in the above equation, we obtain

$$\begin{aligned} \cos\psi &= \frac{(R\cos\delta - r\sin\theta\cos\phi)(\Omega r\sin\theta\sin\phi - \omega R\sin\delta)}{s v_{rel}} \\ &\quad + \frac{(R\sin\delta - r\sin\theta\sin\phi)(\omega R\cos\delta - \Omega r\sin\theta\cos\phi)}{s v_{rel}} \\ &= \frac{rR\sin\theta}{s v_{rel}}\sin(\phi-\delta).(\Omega - \omega). \end{aligned} \tag{8.40}$$

Further, we find that

$$|\cos\psi| = \frac{rR\sin\theta}{s v_{rel}}|\sin(\phi - \delta)||(\Omega - \omega) \tag{8.41}$$

since $\sin\theta \geq 0$ for $0 \leq \theta \leq \pi$.

The torque of the force acting on the satellite (about the spinning axis)due to velocity induction of dm

$$\begin{aligned} dT_z &= (\mathbf{R} \times d\mathbf{F})_z = (\mathbf{R} \times dF\frac{\mathbf{s}}{s})_z \\ &= \frac{dF}{s}(R\cos\delta.s_y - R\sin\delta.s_x). \end{aligned}$$

where s_x and s_y are the x and y components of s. Substituting the expression for dF and simplifying

$$dT_z = -\frac{G_0.dm.m_P}{c^2 s^3} v_{rel}^2 \cos\psi.|\cos\psi| rR\sin\theta\sin(\delta - \phi).$$

Using (8.40) and (8.41) in the above relation and simplifying, we obtain

$$\begin{aligned} dT_z &= -\frac{G_0.dm.m_P}{c^2 s^5} r^3 R^3 \sin^3\theta \sin^2(\delta - \phi). \\ & |\sin(\delta - \phi)|.|\Omega - \omega|(\Omega - \omega), \end{aligned}$$

where

$$dm = \rho(r) r^2 dr \sin\theta d\theta d\phi.$$

Integrating over the whole volume of the planet using the symmetry

$$\begin{aligned} T_z &= -\frac{2G_0 m_P R^3}{c^2} \int_0^{r_p}\int_0^{\pi/2}\int_0^{2\pi} \frac{\rho(r) r^5 \sin^4\theta \sin^2(\delta - \phi).|\sin(\delta - \phi)|}{\{R^2 + r^2 - 2rR\sin\theta.cos(\phi - \delta)\}^{5/2}} \\ & \times d\phi.d\theta.dr \times |\Omega - \omega|(\Omega - \omega). \end{aligned} \tag{8.42}$$

If we take the particular case of Phobos and Mars, the data can be substituted into (8.42) and the integration can be done numerically using a computer. In the case of Phobos $\omega > \Omega$ and, thus, the resultant force on the satellite due to velocity-dependent inertial induction of the planet Mars will be a drag force. This will cause a secular angular acceleration

$$\dot{\omega}_{\text{Ph}} = \frac{T_z}{m_P R^2}.$$

After computation we get

$$\dot{\omega}_{\text{Ph}} = 0.81 \times 10^{-20} \quad \text{rad s}^{-2}.$$

Observations in the past indicated the presence of a secular acceleration of Phobos. The most recent[7] and most dependable observations have detected a secular acceleration of the orbital motion of Phobos. The magnitude is given by

$$(\dot{\omega}_{\text{Ph}})_{obs} \approx 1.05 \times 10^{-20} \quad \text{rad s}^{-2}.$$

The agreement is quite remarkable.

Though the tidal phenomenon (of the solid Mars and its molten core) has been suggested as the source, many have expressed doubt about it.

There is another satellite of Mars, Deimos, which revolves round the planet along a larger orbit. The data for Deimos are as follows:

$$\begin{aligned} R_{De} &= 2.35 \times 10^{7} \quad \text{m} \\ \omega_{De} &= 5.8 \times 10^{-5} \quad \text{rad s}^{-1} \end{aligned}$$

If we carry out a similar analysis, we find that the orbital velocity is subjected to a secular retardation (because $\omega_{De} < \Omega_{\text{Mars}}$) of magnitude

$$4.94 \times 10^{-23} \quad \text{rad s}^{-2}.$$

Unfortunately, the level of accuracy with which the secular change in the orbital motion of Deimos has been measured[8] is not high enough. One reason is that it is much smaller than in the case of Phobos. The most recent observation has detected a secular retardation of

$$2.46 \times 10^{-23} \quad \text{rad s}^{-2},$$

but the standard error of measurement is about three times this value. However, the agreement is reasonable. The above results indicate a reasonable degree of success of the theory of velocity-dependent inertial induction. In future, efforts should be made to detect any secular retardation of the rotation of Mars, and the results would be decisive.

8.7 Transfer of Solar Angular Momentum

Next we turn to a long outstanding problem of astronomy. Although a few alternatives had been proposed, the nebular hypothesis for the origin of the solar system is currently accepted by most scientists. This hypothesis was originally suggested by the German philosopher Emmanuel Kant. However, a major difficulty has forced scientists and cosmologists to look for alternatives. Let us explain in some detail.

[7] Sinclair, A. T., *Astronomy and Astrophysics*, V.220, 1989, p.321.

[8] *Ibid*

According to the nebular hypothesis the stars are formed through gravitational collapse of clouds of gas and dust. Since these dust and gas clouds occupy a very large region of rotating galaxies, such clouds possess angular momentum. Initially the gas and dust cloud collapses under self -gravitation. The gravitational energy is lost through thermal radiation as the system heats up. The angular momentum of the cloud remains constant and, as a result, it rotates at increasingly faster speeds when it shrinks to smaller sizes. When the angular speed is large enough, the material from the equatorial region stops falling in due to gravity. At this stage the centrifugal force in the equatorial region becomes strong enough to prevent further gravitational collapse. This phase spans a period of about 10^6 years. After this, a proto-planetary disk develops out of the material left from the equatorial region of the shrinking cloud. The cloud, however, continues falling inwards due to stronger gravitational, pull and the temperature inside the cloud increases. The efficiency of energy loss decreases due to increased opacity of the denser cloud. After about 10^7 years in typical cases the inside temperature becomes large enough to initiate nuclear fusion. Once the nuclear process starts, the heat generated inside develops enough pressure to arrest any further gravitational collapse of the cloud, and the main sequence phase of the star begins. This period is the longest period and is typically of a duration of about 10^{10} years. The gas and dust in the rotating proto-planetary disk also condense and form a few lumps orbiting the star which ultimately become planets (and their natural satellites). At the onset of the main sequence, the stellar wind generated by the high rate of energy production sweeps the debris away, some of which gets accumulated in the form of a spherical shell shaped cloud enveloping the solar system (it is called the Oort's cloud). The stellar system becomes cleaner, with only a few planets and their satellites orbiting the star. Figure 8.13 shows the stages of star formation in a schematic manner.

The above process suggests that the present angular momentum of the system is derived from the original rotating cloud. Therefore, the amount of angular momentum of a body should be proportional to the matter content of the body. A very strange situation arises so far as the actual distribution of angular momentum is concerned. The sun consists of 99.9% of the mass of the solar system (we consider the sun, the nine planets and their satellites and the asteroid belt as the solar system). But strangely, the sun contributes only 0.5% of the angular momentum of the solar system. This could not be explained by the proponents of the nebular theory. Because of this serious difficulty, the nebular hypothesis remained almost abandoned for a long time. Jeans' theory, based on the hypothesis of ejection of a cigar shaped material from the sun due to a close approach by another star, was put forward to explain the formation of the planets. However, it was shown later that such a process cannot give rise to the formation of planets because the ejected gas is much more likely to dissipate due to the high temperature. The fact that all planets rotate round the sun in the same direction, while the planets lie in the equatorial plane of the sun which also rotates in the same direction as the planets, supports the nebular hypothesis very strongly. The nebular hypothesis was, therefore, revived with a proposal that most of the angular momen-

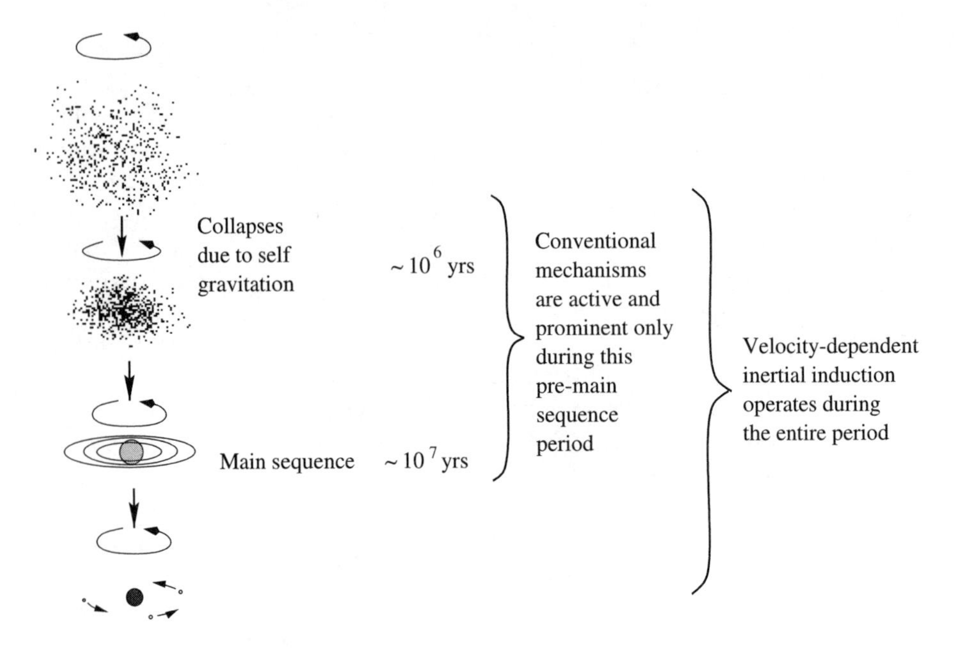

Figure 8.13: Stages in the life of the solar system.

tum of the sun is transferred to the planets. In modern times, the original theory of Kant and Laplace was revised by von Weizsäcker, Kuiper and others, who suggested that turbulence and friction play an important role. However, serious doubt has been expressed about the feasibility of such a process, which could operate only during the short pre-main-sequence period. After the sun enters the main sequence and the gas and dust get swept away by the solar wind, no friction could exist. On the other hand, 'hot' theories have been developed by Alfvén, Hoyle and others in which the ionization of the original cloud and the electromagnetic forces are assumed to be responsible for the transfer of solar angular momentum. Hoyle proposed that the solar wind, which carries a charge, acts like spokes connected to the planets and the ionized material in the proto-planetary disk. These spokes cause the transfer of angular momentum from the spinning sun to the planets. Of course, the intensity of this process at present is totally inadequate for transferring almost all the angular momentum. What has been suggested is that during the transition period the intensity of the solar wind was very strong. However, it is very doubtful if the intensity could have been so high as to transfer almost the whole angular momentum in the relatively short period before the gaseous material in the planetary disk was dissipated.

It has already been shown in Section 8.2 that the proposed velocity dependent inertial induction can transfer angular momentum from a spinning central body to an orbiting body. An approximate analysis will be attempted here because we have to consider the whole evolutionary period. In analyzing the Mars-Phobos system we found that an accurate analysis involves numerical computation, as it is difficult to derive a time dependent equation in a closed form. There are many parameters, such as the original size of the cloud, the fraction of the mass forming the proto-planetary disk, *etc.* which can be estimated from the current physical properties, chemical compositions and the sizes of the planets and comets. However, these estimates cannot be very accurate. At the same time, the characteristics of the transfer of angular momentum depend quite sensitively on the size of the original cloud (and its angular momentum) and the size of the disk which separates out to form the planets and other components of the solar system. Hence, a very accurate mathematical model is neither necessary nor practicable. This section presents an approximate analysis only to demonstrate the possibility that velocity-dependent inertial induction might provide a plausible mechanism of the transfer. It has been found that the effect of the rest of the galaxy (and the universe) is much smaller than local interactions. No attempt will be made here to investigate the formation of the proto-planetary disk and planets. Since the objective is to investigate the approximate order of magnitude of the angular momentum that can be transferred from the sun, we will assume a certain fraction of the original nebula to be detached from the equatorial region of the spinning proto-sun once the stage of rotational instability is reached. Subsequently the proto-sun reaches the main sequence in about 2×10^7 years and the angular momentum is transferred to the detached disk, a part of which later forms the planets. The majority of scientistsaccept that most of the proto-planetary disk material gradually evaporated during the long main-sequence period (especially in the early stages).

It should be remembered that the material which is dislodged from the equatorial region due to rotational instability does not continue to shrink with the rest of the proto-sun. On the other hand, the proto-sun shrinks further and gains angular speed. Thus the central body rotates at a speed higher than that of the detached proto-planetary matter, resulting in a relative motion. Consequently, the proposed mechanism of velocity dependent inertial induction becomes active, and angular momentum of the central spinning body can be transferred to the proto-planetary disk rotating at a relatively slower speed. Figure 8.14 shows a spinning sphere (representing the proto-sun and then the sun in the main-sequence) and another body of mass m in the equatorial plane. It should be noted that whether we treat the body as a particle or a ring with a radius equal to the distance of the particle, the result will be the same. Hence, the proto-planetary ring will be assumed to be a particle at any point on the ring.

First of all, we consider a thin slice of the rotating central body at a distance x from the centre O along the axis of rotation. The thickness of the elemental slice of the sphere is dx. The radius of this elemental disk is $(r_s^2 - x^2)^{1/2}$, where r_s is the radius of the central sphere at a particular point of time. Next we consider an elemental ring of this thin disk of radius r and width dr. Finally, we take an element A of this ring at an angle θ as shown in the figure. Since only the relative motion is responsible for generating inertial induction effect, we consider only the relative spin velocity $\boldsymbol{\Omega}$ of the central body with respect to the separated part. The force on m due to the elemental

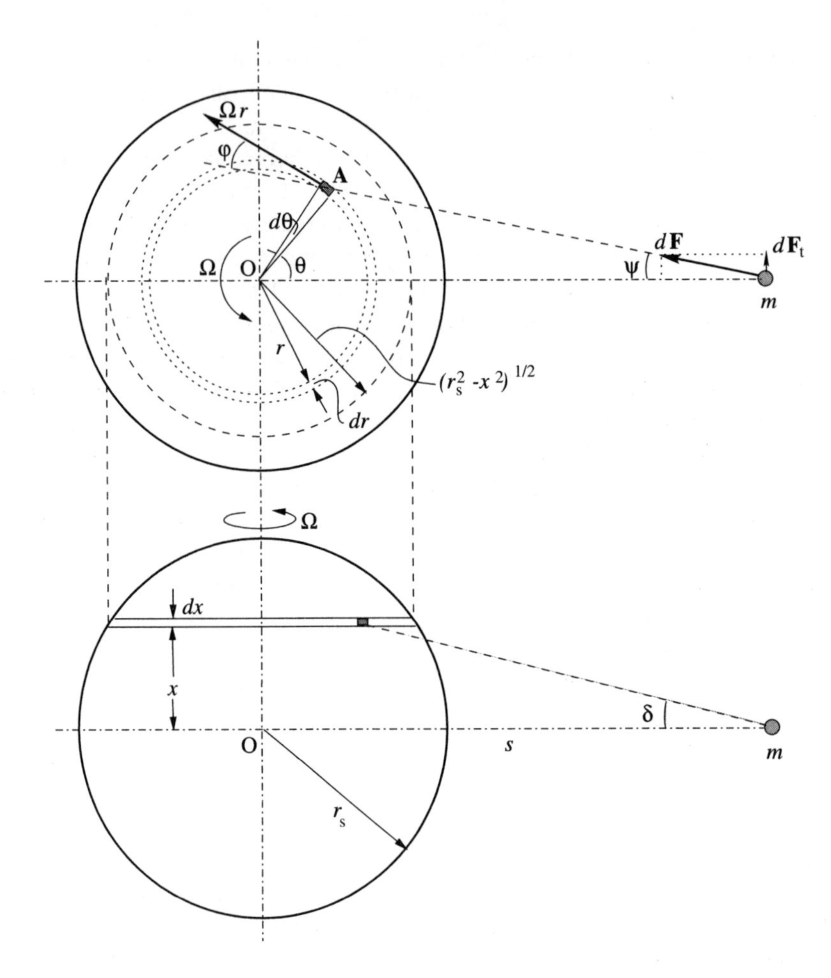

Figure 8.14: Force on a proto-planetary mass due to a rotating central body.

mass A (neglecting the effect of angle δ,,m as will be explained later)[9] is

$$dF \approx \frac{G_0 m \Omega^2}{c^2} \cdot \frac{\rho r \cos^2 \phi}{s^2 + r^2 - 2sr\cos\theta} d\theta dr dx,$$

where ρ is the density (not a constant), Ω is the relative angular speed of the sun[10] and

[9]Ghosh, A., *Earth, Moon and Planets*, V.42, 1988, p.69.

[10]Since the torque is produced due to the difference between the angular velocity of the central body (the proto-sun or sun) and the orbital angular velocity of m, the limiting situation arises when both are equal. Therefore, the minimum distance of the inner edge of the proto-planetary disk (or the nearest planet in the case of a multiple-body planetary system) will be that for which the orbital angular speed is equal to the angular velocity of the solar rotation. This is about 0.3AU or the radius of the planet Mercury. This explains why the innermost planet, Mercury, is placed at its particular distance from the sun.

the other quantities are as shown in the figure. The transverse component of this force is

$$dF_t \approx \frac{G_0 m\Omega^2}{c^2} \cdot \frac{\rho r^3 \cos^2\phi \sin\psi}{s^2 + r^2 - 2sr\cos\theta}\, d\theta dr dx$$

Consequently, the total torque about the sun's centre developed by the velocity-dependent inertial induction of the whole rotating central body (proto-sun and then the sun during the main sequence) is given by

$$T \approx \frac{2G_0 m\Omega^2}{c^2} \int_0^{r_s} \int_0^{\sqrt{r_s^2 - x^2}} \rho r^3 s \left(\int_0^{\pi} \frac{\cos^2\phi \sin\psi}{s^2 + r^2 - 2sr\cos\theta} d\theta \right) dr dx, \qquad (8.43)$$

where r_s is the outer radius of the spinning central body. Since, during most of the period, s will be much larger than r_s, the following simplifying assumptions are made without introducing too great an error:

$$\delta \approx 0$$

$$\cos^2\phi \approx \sin^2(\theta + \psi) \approx \sin^2\theta$$

$$\sin\psi \approx (r/s)\sin\theta,$$

and

$$s^2 + r^2 - 2sr\cos\theta \approx s^2. \qquad (8.44)$$

With these simplifications (8.43) becomes

$$T \sim \frac{4G_0 m\Omega^2}{3c^2 s^2} \int_0^{r_s} \int_0^{\sqrt{r_s^2 - x^2}} \rho r^4 dr dx. \qquad (8.45)$$

The density ρ of the proto-sun and sun varies quite significantly with the distance from the centre. The pattern of variation in each case of the proto-sun and the sun is also not the same. During the pre-main-sequence period, the moment of inertia of the central body can be expressed as:

$$I_s \approx 0.1 m_s r_s^2 \qquad (8.46)$$

as suggested by Hoyle.[11] According to him, the sun's moment of inertia in the main-sequence phase can be expressed as

$$I_s \approx 0.05 m_s r_s^2, \qquad (8.47)$$

[11] Hoyle, F., *Quart. Jr. Royal Astronomical Society*, v.1., 1960, p.28.

where m_s is the total mass of the central body. For the sake of simplicity the density variation within the sun can be expressed approximately as follows using the available data:[12]

$$\rho \approx \rho_0 \exp(-8a/r_s), \tag{8.48}$$

where $a(=\sqrt{r^2+x^2})$ is the distance of the element from the centre O. Since the pre-main-sequence period is relatively short, and, according to the proposed theory, does not contribute significantly to the transfer of angular momentum, the density function given by (8.48) will be used for further analysis. Using (8.48) in (8.45) we get

$$T \sim 6 \times 10^{-3} \frac{G_0 m_s \Omega^2 r_s^2}{c^2 s^2}. \tag{8.49}$$

Let L be the total angular momentum of the original cloud, and the ratio of the masses of the sun and the ejected body be given by

$$\frac{m}{m_s} = f. \tag{8.50}$$

If r'_s is the radius of the central body when rotational instability occurs, and the mass is dislodged from the equatorial region, the total angular momentum (same as that of the original cloud) can be expressed as

$$\begin{aligned} L &\approx 0.1 m_s r_s'^2.\Omega' + m r_s'^2 \Omega' \\ &\approx l_{s_0} + m r_s'^2 \Omega', \end{aligned}$$

where l_{s_0} is the angular momentum at the start of the process of transfer. Using (8.50) in the above equation

$$\begin{aligned} L &\approx l_{s_0} + f.m_s r_s'^2 \Omega' \\ &\approx l_{s_0} + 10 f l_{s_0}, \end{aligned}$$

as

$$l_{s_0} = 0.1 m_s r_s'^2 \Omega'$$

Hence

$$l_{s_0} \approx L/(1+10f). \tag{8.51}$$

If l_s is the angular momentum at a time t since the onset of the angular momentum transfer process, then

$$l_s \approx 0.05 m_s r_s^2 \Omega \tag{8.52}$$

[12] Priest, E.R., *Solar Magnetohydrodynamics*, D. Reidel Pub.Co., Dordrecht, Holland, 1982.

because of (8.47). Now at any instant the angular momentum is approximately equal (neglecting other losses) to the difference between the original angular momentum of the cloud, L, and l_s. This is because the total angular momentum of the system must be approximately conserved. So,

$$l_s \approx 0.05 m_s r_s^2 \Omega \tag{8.53}$$

because of (8.47). Now at any instant the angular momentum is approximately equal (neglecting other losses) to the difference between the original angular momentum of the cloud, L, and l_s. This is because the total angular momentum of the system must be approximately conserved. Thus,

$$m s^2 \omega \approx (L - l_s), \tag{8.54}$$

where ω is the orbital angular speed of the dislodged material at the instant considered. We also know

$$m\omega^2 . s = \frac{G_0 m_s m}{s^2},$$

or,

$$\omega^2 = \frac{G_0 m_s}{s^3}. \tag{8.55}$$

Squaring both sides of (8.54) and using (8.55)

$$\frac{G_0 m_s}{s^3} \approx \frac{(L - l_s)^2}{m^2 s^4}.$$

Hence

$$s \approx \frac{(L - l_s)^2}{G_0 m_s m^2}. \tag{8.56}$$

If l_{p_0} be the starting momentum of the dislodged body [$= m r_s'^2 \omega' = 10\, fL/(1 + 10\, f)$], then

$$L = l_{s_0} + l_{p_0}$$

and (8.56) can be written as follows:

$$s \approx \frac{(l_{s_0} + l_{p_0} - l_s)^2}{G_0 m_s m^2}. \tag{8.57}$$

Now, T is equal to the rate of increase of angular momentum of m, or the rate at which the solar angular momentum decreases. Hence, replacing Ω and s from (8.55) and (8.57) in (8.49) we have

$$\frac{dl_s}{dt} \approx -2.4 \frac{G_0^3 m_s m^5}{c^2} \frac{l_s^2}{r_s (L - l_s)^4}. \tag{8.58}$$

The negative sign is because of the decrease in l_s. Since r_s reduces to about the present solar radius in approximately 2×10^7 years, *i.e.*, during the pre-main-sequence period, this variation in r_s is incorporated by using

$$r_s \approx r_s^p \left[1 + \lambda \exp(-\alpha t)\right],$$

where $(1 + \lambda) r_s^p$ is the radius of the sun when it enters the main-sequence (this is not much different from the present value, r_s^p) and $\lambda \approx 1.35 \times 10^4$ (when t is expressed in seconds). Using the above expression for r_s in (8.58) we finally get

$$\frac{dl_s}{dt} \approx -2.4 \frac{G_0^3 m_s m^5}{c^2} \cdot \frac{l_s^2}{r_s^p [1 + \lambda \exp(-\alpha t)](L - l_s)^4} . \tag{8.59}$$

It should be remembered that, as suggested by Hoyle, a large proportion of the proto-planetary disk dissipates. So m cannot be treated as a constant. But, the analysis becomes very complicated and intractable if the variation in m is considered. Moreover, the information regarding m as a function of time is not available. To keep this very approximate analysis tractable, m has been assumed to be constant at an average value. Now solving (8.59) and using the initial conditions we get

$$\begin{aligned}
&\mu t + \frac{\mu}{\alpha} \ln\left[\{1 + \lambda \exp(-\alpha t)\}/(1 + \lambda)\right] \approx \\
&(10f + 1)^4 (l_{so}^4 / l_s) - 4(10f + 1)^3 l_{so}^3 \ln(l_{so}/l_s) \\
&-6(10f + 1)^2 l_{so}^2 l_s + 2(10f + 1) l_{so} l_s^2 - l_s^3/3 \\
&+\{6(10f + 1)^2 + 1/3 - (10f + 1)^4 - 2(10f + 1)\} l_{so}^3,
\end{aligned} \tag{8.60}$$

where

$$\mu = 2.4 \frac{G_0^3 m_s m^5}{c^2 r_s^p} = 2.4 \frac{G_0^3 m_s^6 f^5}{c^2 r_s^p} .$$

It can easily be seen that when $t >> 2 \times 10^7$ years, the contribution of the second term on the LHS of (8.46) will be negligible compared to the first term. Similarly when $l_{s_o}/l_s >> 1$ the major contribution comes from the first term in the RHS of (8.60). Therefore, for the order of magnitude calculation we can use

$$\mu t \sim (10f + 1)^4 l_{s0}^4 / l_s = L^4 / l_s, \tag{8.61}$$

where $t >> 2 \times 10^7$ years and $l_{s0}/l_s >> 1$.

The condition of the rotational instability of the proto-sun is given by

$$r_s'^3 \Omega'^2 = G_0 m_s$$

or,

$$r_s'^2 \Omega' = (G_0 m_s r_s')^{1/2} .$$

Using this condition, and noting that after the separation of the disk from the central body, the angular momentum of the proto-sun is given by

$$l_{s_0} = L/(1+10f).$$

As shown earlier, the radius of the proto-sun (which is also equal to the inner radius of the disk at this instant) can be expressed as

$$\begin{aligned} r'_s &= 100 l_{s_0}^2 / G_0 m_s^3 \\ &\approx 100 L^2 / \{1+10f)^2 G_0 m_s^3, \end{aligned} \tag{8.62}$$

because

$$l_{s0} \approx 0.1 m_s r_s'^2 \Omega',$$

where the primed quantities refer to those satisfying the rotational instability condition.

Now (8.49) can be used to find out the rate at which the angular momentum of an orbiting mass increases by replacing m by Δm. However, Ω changes continuously depending on the total interaction with the whole disk (or planetary system). For an approximate order of magnitude analysis we take the average value of Ω^2 and treat it as a constant. Moreover, we take $r'_s = r^p_s$, since the sun has been in the main sequence for most of the intervening period and the size has not changed much. With these assumptions (8.49) can be easily solved to yield s as a function of time.

The angular momentum (l_p) of an orbiting mass Δm, can be expressed as

$$l_{\Delta p} = \Delta m s_\Delta^2 \omega,$$

where s_Δ is the distance of Δm from O.

Again, to satisfy Kepler's law, $\omega^2 = G_0 m_s / s_\Delta^2$. Hence

$$l_{\Delta p} = \Delta m (G_0 m_s s_\Delta)^{1/2}.$$

The rate of increase will be equal to T, as mentioned. So

$$\frac{dl_{\Delta p}}{dt} = T.$$

Or,

$$\frac{1}{2} \Delta m \frac{\sqrt{G_0 m_s}}{s_\Delta^{1/2}} \dot{s_\Delta} = 6 \times 10^{-3} \frac{G_0 m_s \Delta m}{c^2 s_\Delta^2} (\Omega)^2_{av} r_s^{p3}$$

if (8.49) is used with Ω_{av}. Solving the above differential equation and simplifying, we get

$$s_{\Delta}^{5/2} \approx \frac{3\times 10^{-2}\sqrt{G_0 m_s}}{c^2}(\Omega)^2_{av} r_s^{p3}.t + A \tag{8.63}$$

where A is a constant. From (8.62) the starting value of s_{Δ} (which will be equal to r'_s) can be found, which will give the value of A in (8.63). After about 4.7×10^9 years (the present age of the solar system) the current value of the inner radius of the planetary system can be obtained from (8.63).

No definite value of L is known to us, but as per Hoyle's estimate it is about 4×10^{44} kg m$^2 s^{-1}$. The estimate has been made by considering the characteristics of the clouds which give rise to stars on collapse. Hoyle further estimated that about 10% of the original cloud formed the proto-planetary disk. A major fraction of this disk dissipated during the long main-sequence period, leaving only a small fraction in the form of the planetary system. This implies that although 99% of the disk has blown away, it took only about 90% of the disk's angular momentum, leaving about 3×10^{43} kg m$^2 s^{-1}$ for the existing planetary system. Similar proposals were made earlier by Weizsäcker.[13]

If we assume $L \sim 4\times 10^{44}$ kg m$^2 s^{-1}$ and $m \sim 0.023 m_s$ (*i.e.*, about 2.3% of the original cloud), substituting $t = 4.7\times 10^9$ years, the present angular momentum of the sun becomes

$$l_s \sim L^4/\mu t \sim 1.4\times 10^{41}\text{kg m}^2 s^{-1},$$

which agrees nicely with the present observed value 1.5×10^{41}kg m$^2 s^{-1}$. Most of the original angular momentum is transferred to the planetary disk, which evaporates (as per Hoyle and Weizsäcker) leaving behind only 4% of the disk mass and 30% of the momentum to constitute the present planetary system.

Taking $L \sim 4\times 10^{44}$kg m^{2}s^{-1} as suggested by Hoyle, it can be shown that the present value of the solar angular momentum is obtained after 4.7 billion years if f is taken as 0.07. Using these values of L and f, the radius at which the separation of matter from the proto-sun ends (or the initial value of the disk's inner radius) is found from (8.62) to be 10.38×10^9m. The current radius of the innermost particle Δm can be found using (8.63) to be about 2.1×10^{10} m, which is close to the orbital radius of Mercury — again a very satisfactory result. It was shown earlier that at this radius the velocity-dependent inertial induction is not present, and the smallest orbit will be given by the condition that the relative rotation vanishes.

It should be clear that these values are very approximate, and the analysis only demonstrates the feasibility of the proposed velocity-dependent inertial induction as a mechanism to transfer the solar momentum. There is one major difference between this model and the previous theories. In the conventional theories proposed earlier by Weizäcker and Hoyle, most of the transfer of angular momentum took place in the relatively short pre-main-sequence period. On the other hand, according to the proposed

[13] von Weizsäcker, C.F., *Zeit f. Astrophysik*, V.22, 1943, p.319; von Weizsäcker, C.F., *Zeit f. Astrophysik*, V.24, 1947, p.181.

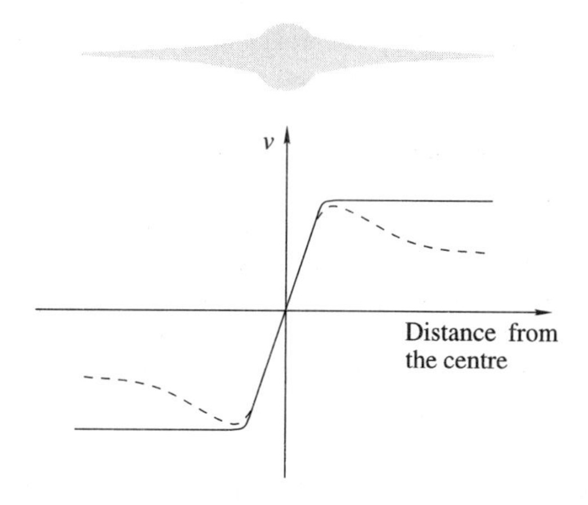

Figure 8.15: Characteristics of orbiting speed of stars in a spiral galaxy.

mechanism based on velocity-dependent inertial induction, most of the transfer takes place during the long main-sequence period. This implies that any star which has not gone far into the main-sequence period (*i.e.*, which has recently finished the pre-main sequence period) still preserves most of its original angular momentum. Or, a new born star should generally rotate faster, and very old stars should be slow rotators. This is actually found to be so by a large number of observations. The conventional theories, however, cannot explain this feature, as most of the transfer is supposed to be completed soon after the main-sequence starts.

In reality, in the pre-main sequence period, the transfer may take place through more than one mechanism in which inertial induction plays a major role. During the main-sequence period, however, it is the only factor.

8.8 Servomechanism for Matter Distribution in Spiral Galaxies

Another very interesting effect of velocity-dependent inertial induction is that it can explain a curious phenomenon in all spiral galaxies, as already mentioned in Chapter 2. It has now been firmly established that almost all spiral galaxies have flat rotation characteristics. The stars in these galaxies have almost circular orbits. The most startling fact is revealed when the velocities of these stars are measured. Figure 8.15 shows the plot of the orbital velocity with the distance from the galactic centre. The central region rotates almost like a rigid body and the orbital speed increases linearly with distance. But the stars in the spiral arms rotate around the centre with almost constant velocity, as indicated in Fig. 8.15. If the luminosity is assumed to represent the matter present (in the form of stars), then according to Newtonian gravitation theory the velocity curve should gradually fall off as shown by the broken line. Some have sought to explain the conspicuous absence of the Keplerian fall-off by the hypothesis that a large proportion of the galactic matter is dark.[14] Milgrom has attempted to explain the phenomenon by modifying Newton's laws, as already mentioned in Chapter 2. Whatever the case may be, a crucial point identified by a number of researchers is that a flat rotation curve requires a unique distribution of matter in the galactic disk. The unique distribution of matter required for a flat rotation curve cannot be obtained in the galaxies by chance. Since the flat rotation curve is almost a universal feature in all spiral galaxies, a servomechanism must exist to distribute the matter in the unique way. Until now, no such servomechanism has been reported using conventional physics. This section presents a plausible servomechanism based on the proposed theory of velocity-dependent inertial induction.[15]

In this analysis, the spiral galaxies will be assumed to be axisymmetric thin disks truncated at a suitable radius within which most of the galactic matter is contained. This obviously idealizes the situation; but the main objective of this analysis is only to indicate the feasibility of inertial induction to provide the necessary servomechanism.

As in all other cases of this chapter, the effect of the universal interaction due to velocity will be neglected, as the local interaction predominates. In the conventional Newtonian approach, the contribution of the first term of (4.2), the usual gravitational pull towards the centre, and that from the third term (the centrifugal force) are made to neutralize so that an equilibrium is achieved. However, the solution is not unique. When the velocity drag (the contribution of the second term) is considered, another relationship between the mass distribution and the velocity profile is obtained, so that the total tangential force on a mass particle is also zero. If the resultant tangential force on a particle acts in the direction of the orbital motion (call it a pull), the particle will move away from the centre (this happens in all gravitating objects). As the particle moves away from the centre, the pull will also gradually decrease until it becomes zero and an equilibrium is reached. Similarly, if the tangential force on a particle is a drag, it will lose angular momentum and move nearer to the centre until the tangential force becomes zero and an equilibrium is reached. If a particle which is in equilibrium is pushed out, it will be subjected to a drag that will cause the particle to again return to its equilibrium orbit. On the other hand, if it is pushed inwards, an accelerating force will develop causing the particle to move out, and again attain the equilibrium orbit. Thus, we find that velocity-dependent inertial induction not only distributes the matter in a unique way so that each particle is subjected to zero radial force and zero tangential force, but it also ensures that this equilibrium is a stable one. When both the radial and transverse equilibrium conditions are considered, unique solutions for both the mass distribution and velocity profile are obtained.

Figures 8.16(a) and 8.16(b) show the interaction configuration of a point mass S

[14]Toomre, A., *Astrophysical Jr.*, V.138, 1963, p.385; Freeman, K. C., *Astrophysical Jr.,*, V.160, 1970, p.811.

[15]Ghosh, A., Rai, S and Gupta, A., *Astrophysics and Space Science*, V.141, 1988, p.1.

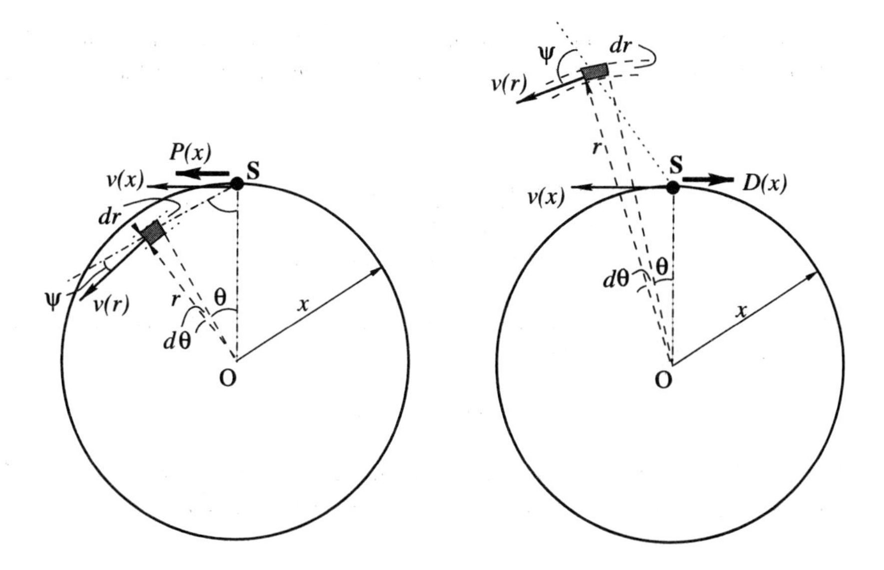

Figure 8.16: Pull and drag force on a star due to differential rotation of the galaxy.

with an elemental mass inside and outside the orbit of S, respectively. The resultant of the velocity-dependent interaction of S with all the matter present inside the orbit is a pull $P(x)$, as shown. All matter inside the orbit has a higher angular speed than S, and so the induction results in a pull. On the other hand, the angular speed of any matter outside the orbit of S is lower than that of S and, therefore, the inertial induction results in a drag. The resultant drag is $D(x)$ as shown in Fig. 8.16(b).

In this analysis, the variation of G with the distance between the interacting particles is ignored. Since $G = G_0 \exp(-4 \times 10^{-27} r)$, the change becomes perceptible only when the distance involved is much greater than the size of a typical spiral galaxy. Once again, all the objects are assumed to follow circular orbits.

When the radius of $S's$ orbit is x, mass of S is m, the surface density of matter in the disk is $\sigma(r)$ and the velocity profile is $v(r)$, the pull on S due to the matter present inside the orbit of S can be expressed as[16]

$$P(x) = \frac{2G_0 m}{c^2} \int_0^x \int_0^\pi \frac{\sigma(r) r \{v(r)\cos\psi - v(x)\sin\phi\} |v(r)\cos\psi - v(x)\sin\phi|}{r^2 + x^2 - 2rx\cos\theta} \times \sin\phi . d\theta dr. \quad (8.64)$$

[16]Ghosh, A. *et al.*, *Astrophysics and Space Science*, V.141, 1988, p.1.

The above expression is obtained by integrating the tangential component of the force developed by an element of the disk, which can be expressed in the following form:

$$
\begin{aligned}
dP(x) &= \frac{G_0 m \sigma(r) r d\theta dr}{c^2(r^2 + x^2 - 2rx\cos\theta)}\{v(r)\cos\psi - v(x)\sin\phi\} \\
&\quad \times |v(r)\cos\psi - v(x)\sin\phi| \cdot \sin\phi. \qquad (8.65)
\end{aligned}
$$

It should be noted that the $v^2 \cos\theta \cdot |\cos\theta|$ term in the expression for velocity-dependent inertial induction is equivalent to

$$\{v(r)\cos\psi - v(x)\sin\phi\}|v(r)\cos\psi - v(x)\sin\phi|.$$

If we assume the galactic disk to be truncated at a radius R, and use the nondimensional quantities $\xi = r/R$ and $\eta = x/R$, then (8.64) can be rewritten in the form

$$
\begin{aligned}
P(\eta) &= \frac{2G_0 m}{c^2}\int_0^{\eta}\int_0^{\pi} \\
&\quad \frac{\sigma(\xi)\xi\{v(\xi)\cos\psi - v(\eta)\sin\phi\}|v(\xi)\cos\psi - v(\eta)\sin\phi|\sin\phi d\theta d\xi}{\xi^2 + \eta^2 - 2\xi\eta\cos\theta}.
\end{aligned}
$$

Integrating over θ, we obtain

$$
\begin{aligned}
P(\eta) &= \frac{8G_0 m}{3c^2}\int_0^{\eta} \\
&\quad \frac{\xi^2}{\eta^3(\eta^2 - \xi^2)}\sigma(\xi)\{\eta v(\xi) - \xi v(\eta)\}|\eta v(\xi) - \xi v(\eta)|d\xi. \qquad (8.66)
\end{aligned}
$$

The expression for the drag on S due to the matter present in the disk outside the orbit of S can be determined in a similar way:

$$
\begin{aligned}
D(x) &= \frac{2G_0 m}{c^2}\int_x^{R}\int_0^{\pi} \\
&\quad \frac{\sigma(r) r\{v(x)\sin\phi - v(r)\cos\psi\}|v(x)\sin\phi - v(r)\cos\psi|\sin\phi d\theta dr}{r^2 + x^2 - 2rx\cos\theta}.
\end{aligned}
$$

If we use the same non-dimensional terms $\xi = r/R$, and $\eta = x/R$, the above equation (after integration over θ) becomes

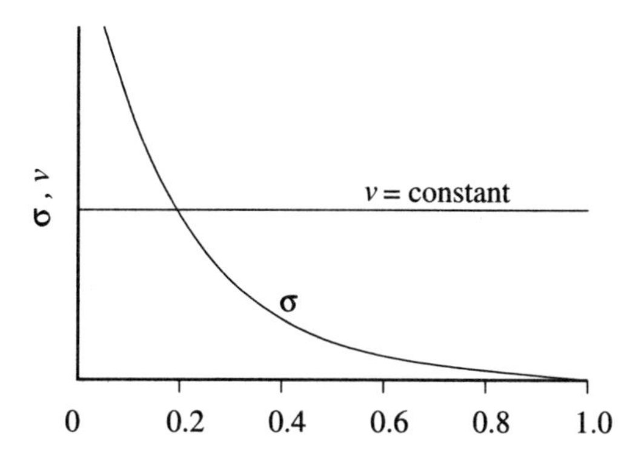

Figure 8.17: Density distribution for a constant velocity situation.

$$
\begin{aligned}
D(\eta) &= \frac{8G_0 m}{3c^2} \int_{\eta}^{1} \\
&\quad \frac{1}{\xi(\xi^2 - \eta^2)} \sigma(\xi)\{\xi v(\eta) - \eta v(\xi)\}|\xi v(\eta) - \eta v(\xi)| d\xi. \qquad (8.67)
\end{aligned}
$$

Equilibrium is achieved when $P(\eta) = D(\eta)$, and the centripetal acceleration is equal to the resultant acceleration due to gravity. A direct solution of these equations in analytical form is impossible. Therefore, an indirect approach is adopted. We consider the velocity profiles to be as observed in the case of spiral galaxies, and show that there is a tendency for $P(\eta)$ to match $D(\eta)$ for all values of η. Exact matching of $P(\eta)$ with $D(\eta)$ is, of course, not expected, because of the approximate nature of the analysis.

To begin with, we take the ideal situation where orbital speed is constant. In such cases we find the following relations:[17]

$$
\begin{aligned}
v(r) &= v_0 = (\pi G_0 M/2R)^{1/2}, \\
\sigma(r) &= (M/2\pi r R)\sin^{-1}(1 - r^2/R^2)^{1/2} \\
&= (M/2\pi R^2)\{\sin^{-1}(1 - \xi^2)^{1/2}\}/\xi,
\end{aligned}
$$

where M is the mass of the disk. The density function σ is derived from the consideration that the acceleration of a particle due to the gravitational pull towards the centre is equal to the centripetal acceleration of the particle due to its orbital motion. Figure 8.17 shows the variation of v and σ.

Substituting the above expressions for v and σ in (8.66) we get

[17] Friedman, A. M. and Polyachenko, V. L., *Physics of Gravitating Systems I, Equilibrium and Stability*, Springer-Verlag, New York, 1984, p.327.

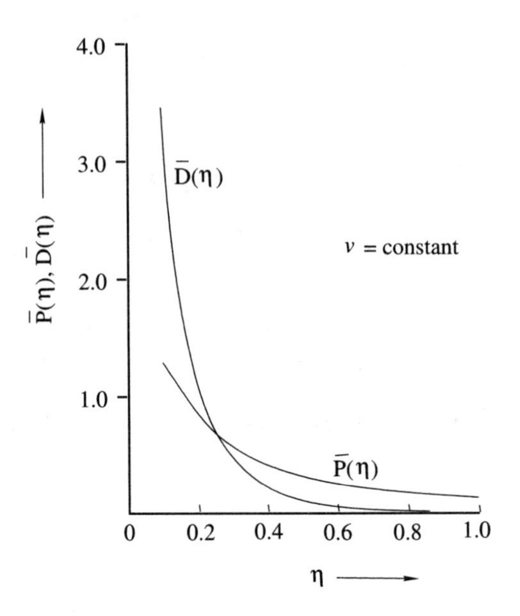

Figure 8.18: Comparison of drag and pull at different radii for velocity distribution shown in Fig.8.17.

$$\begin{aligned} P(\eta) &= \frac{2G_0Mmv_0^2}{3\pi c^2R^2}\left[\frac{1}{\eta^3}\int_0^\eta \frac{\xi(\eta-\xi)}{\eta+\xi}\sin^{-1}\sqrt{1-\xi^2}d\xi\right] \\ &= \frac{2G_0Mmv_0^2}{3\pi c^2R^2}\bar{P}(\eta). \end{aligned} \tag{8.68}$$

Similarly, the expression for $D(\eta)$ becomes

$$\begin{aligned} D(\eta) &= \frac{2G_0Mmv_0^2}{3\pi c^2R^2}\left[\int_\eta^1 \frac{\xi-\eta}{\eta^2(\xi+\eta)}\sin^{-1}\sqrt{1-\xi^2}d\xi\right] \\ &= \frac{2G_0Mmv_0^2}{3\pi c^2R^2}\bar{D}(\eta). \end{aligned} \tag{8.69}$$

When $\bar{P}(\eta)$ and $\bar{D}(\eta)$ are computed numerically, we obtain Fig. 8.18 which shows the variations of $\bar{P}(\eta)$ and $\bar{D}(\eta)$ in the range $0 \leq \eta \leq 1$. Figure (8.18) shows that, although the condition $\bar{P}(\eta) = \bar{D}(\eta)$ is not exactly satisfied in the whole range, the match is reasonably good in the mid-disc region. The speed near the edge and the centre cannot remain constant (and does not remain constant in actual cases). Moreover,

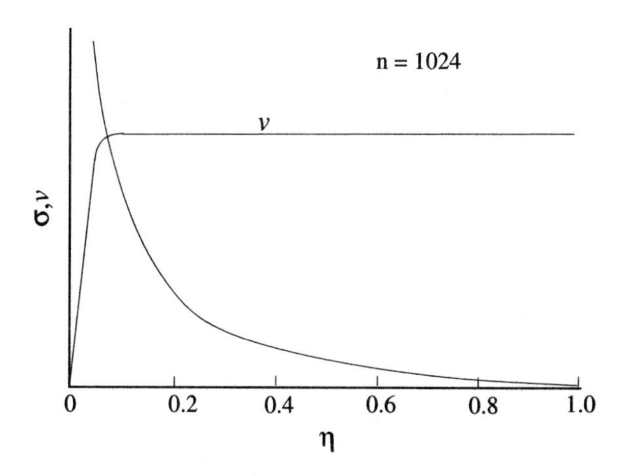

Figure 8.19: A realistic density and velocity distribution.

the model of a thin disc does not fit near the galactic centre either. Thus, the mismatch is expected in the central region and the edge.

Next we can consider another ideal situation in which the orbital speed increases linearly with the radius in the central region, and then remains more or less constant, as shown in Fig. 8.19. This is more realistic and resembles the actual characteristics more closely. The corresponding density function shown in Fig.8.19 can again be determined by equating the centripetal acceleration of a particle to the acceleration due to the gravitational pull on the particle. The orbital speed and the corresponding density functions are given[18] by the following expression:

$$v^2(\xi) = \frac{2n+1}{4n}\left(\frac{\pi G_0 M}{R}\right)[1-(1-\xi^2)^n], \tag{8.70}$$

$$\sigma(\xi) = \frac{M}{2\pi R^2}\sum_{k=1}^{n} b_{k,n}(1-\xi^2)^{k-l/2}, \tag{8.71}$$

where

$$b_{1,n} = \frac{2n+1}{2n-1}; \quad b_{k,n} = \frac{4(k-1)(n-k+1)}{(2k-1)(2n-2k+1)} b_{k-1,n}.$$

The value of the parameter n determines the shape of the velocity curve. When n tends to infinity, the case approaches the ideal situation of constant velocity. Using $v(\xi)$ and $\sigma(\xi)$ from (8.70) and (8.71) in (8.66) and (8.67) we can evaluate the pull and the drag by numerical computation. Taking $n = 1024$ it is seen that $\bar{P}(\eta)$ and $\bar{D}(\eta)$ are sufficiently close, as indicated in Fig.8.20.

[18]Friedman, A.M. and Polyachenko, V.L., *Physics of Gravitating Systems, I, Equilibrium and Stability*, Springer-Verlag, New York, 1984, p.327.

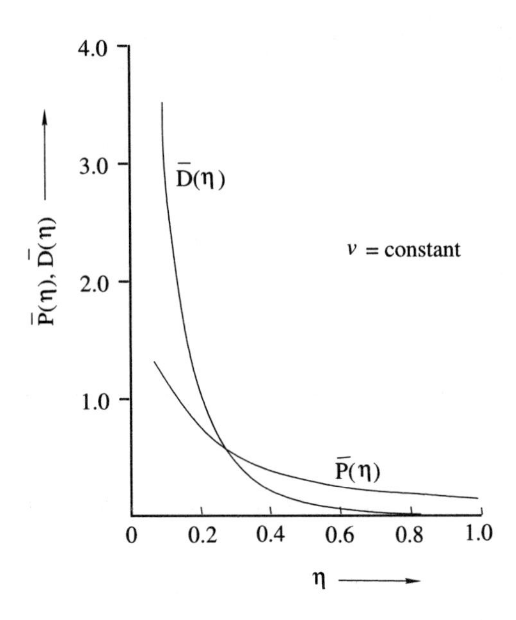

Figure 8.20: Comparison of drag and pull for velocity distribution shown in Fig. 8.19.

At this point it should be mentioned that the $\bar{P}(\eta)$ and $\bar{D}(\eta)$ curves are very sensitively dependent on $v(r)$ and $\sigma(r)$. It has been found that in general the magnitudes of $\bar{P}(\eta)$ and $\bar{D}(\eta)$ are widely different. Only for the cases of $v(r)$ characteristics normally observed in spiral galaxies do $\bar{P}(\eta)$ and $\bar{D}(\eta)$ tend to become similar, as indicated in the figures. As a result, by reverse logic it can be stated that velocity-dependent inertial induction provides a servomechanism for distributing the matter in a unique manner based on the condition that $\bar{P}(\eta) = \bar{D}(\eta)$. This unique matter distribution results in an orbital speed characterized by a flat rotation curve. The above analysis is, no doubt, a very approximate one, but it strongly indicates the possibility that velocity-dependent inertial induction might provide the necessary mechanism. This also helps to resolve the mystery why a flat rotation curve is observed in the case of most spiral galaxies. In conventional physics, no servomechanism have been identified so far.

In this chapter a number of different cases has been considered where object to object velocity-dependent inertial induction can have detectable effects. It is clearly established that in all these cases the observational results strongly support the proposed theory. In most cases, the model resolves the unsolved mysteries and unexplained features. Lastly, it should be noted that these phenomena are unconnected, except for the presence of inertial induction. There are quite a few other problems which need explanation outside conventional mechanics, and, we are hopeful that the proposed model will be able to resolve these issues in future research work.

Chapter 9

Extra-galactic Phenomena

9.1 Introduction

In the previous chapters, a number of cases have been investigated in which it is possible to predict measurable effects produced by the proposed velocity-dependent inertial induction. In all cases, not only are the predicted effects found to be present, but the model also resolves a number of unsolved questions. In all these unrelated phenomena, the quantitative agreement of the predictions with observations warrants a reasonable degree of confidence as to the correctness of the proposed theory. The quantitative agreement is all the more significant because the model does not rely on any freely adjustable parameter. In this chapter a few other phenomena will be presented which are amenable to analysis with the proposed theory of inertial induction.

9.2 True Velocity Dispersion of Galaxy Clusters and the Dark Matter Problem

Dynamical studies of galaxy clusters, assuming the observed redshifts of the individual galaxies to be of purely Doppler origin, yield a very strange result. It is found that the velocities of the galaxies in a cluster are very large, and to keep the cluster together as a gravitationally bound system the total mass of the system has to be much larger than expected from luminous matter. Or in other words, a very large amount of gravitating matter must be in dark form. Early estimates of cluster masses were based on the application of the virial theorem. If it is assumed that a cluster of galaxies is a self-gravitating bound system, then the virial mass is given by

$$M_v \sim 3\frac{R_G v_d^2}{G}.$$

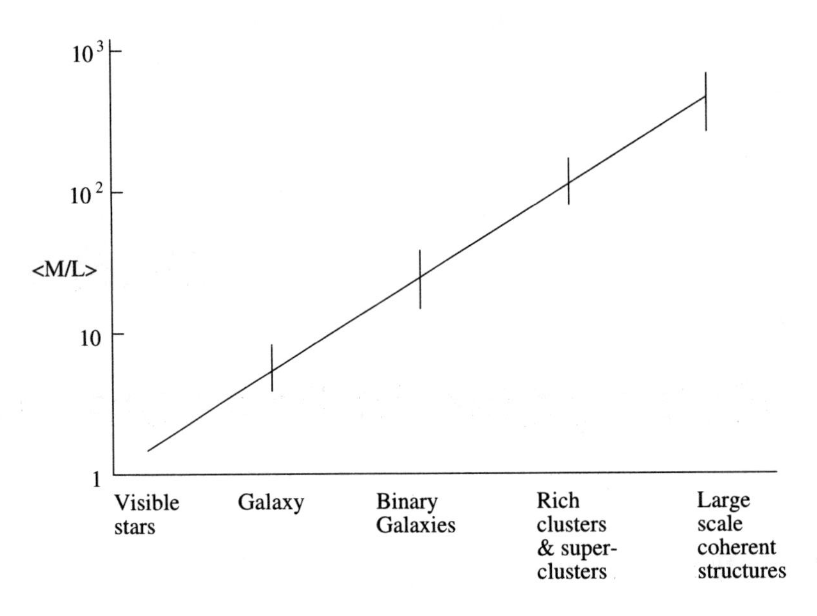

Figure 9.1: Proportion of dark matter with increasing system size.

where M_v is the virial mass, R_G is the gravitational radius and v_d is the velocity dispersion (to be explained in greater detail later in this section).

This has led to the interpretation that most of the universe is composed of dark matter (DM). It is also found[1] that the ratio of dark matter to luminous matter must increase with the size of the system considered, as shown in Fig. 9.1. This raises doubts as to the correctness of the hypothesis that the redshifts of the galaxies are of purely Doppler origin. If a large fraction of the observed redshift is an indicator (in the presence of the proposed cosmic drag mechanism to cause the redshifting of photons) of the distance rather than the recessional velocity of a galaxy along the line-of-sight, then it is obvious that the redshift dispersion indicates the diameter of the system of galaxies under consideration[2]. This explains why the proportion of dark matter increases with the size of the system when the virial theorem is employed to determine the system mass, assuming the redshifts to be purely Doppler. On the other hand, an increase of the ratio of dark matter to luminous matter with the size of the system is difficult to explain, whatever the nature of dark matter distribution. Furthermore, the magnitude-redshift relationship shows that galaxies of higher magnitude possess higher redshifts. Thus, it is reasonable to assume that a major fraction of the redshift is an indicator of

[1] Ostriker, J.P., and Peebles, P.J.E., *Astrophysical Jr.*, V.186, 1973, p.467. Trimble, V., "Existence and Nature of Dark Matter in the Universe" in: *Annual Review of Astronomy and Astrophysics*, V.25, 1987, p.425.

[2] Ghosh, A., *Astrophysics and Space Science*, V.227, p.41, 1995.

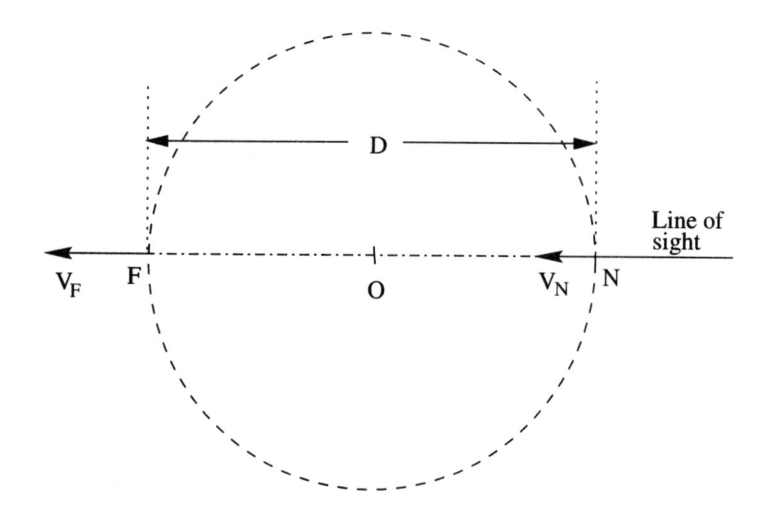

Figure 9.2: Recessional velocities of galaxies at the diametrically opposite ends of a spherical cluster.

distance rather than velocity.

Figure 9.2 shows a system of galaxies (assumed to be approximately spherical, with a diameter D). The nearest and farthest members along the line-of-sight are N and F, respectively. If the redshifts of N and F are z_N and z_F, respectively, then assuming these redshifts to be purely due to recession, the corresponding velocities are as follows:

$$v_N = cz_N,$$

$$v_F = cz_F.$$

As a result, the dynamical configuration with respect to a frame moving with its centre O is as indicated in Fig. 9.2. The velocity of the frame co-moving with the C.G. of the system is

$$v_0 = \frac{v_F + v_N}{2}. \tag{9.1}$$

The velocity dispersion (*i.e.*, the velocity of the outer galaxies with respect to the cluster C.G.) is given by

$$v_d = \frac{v_F - v_N}{2}. \tag{9.2}$$

On the other hand, if the greater part of the redshifts z_N and z_F are due to the cosmic drag, then the actual velocities of the cluster members with respect to the co-

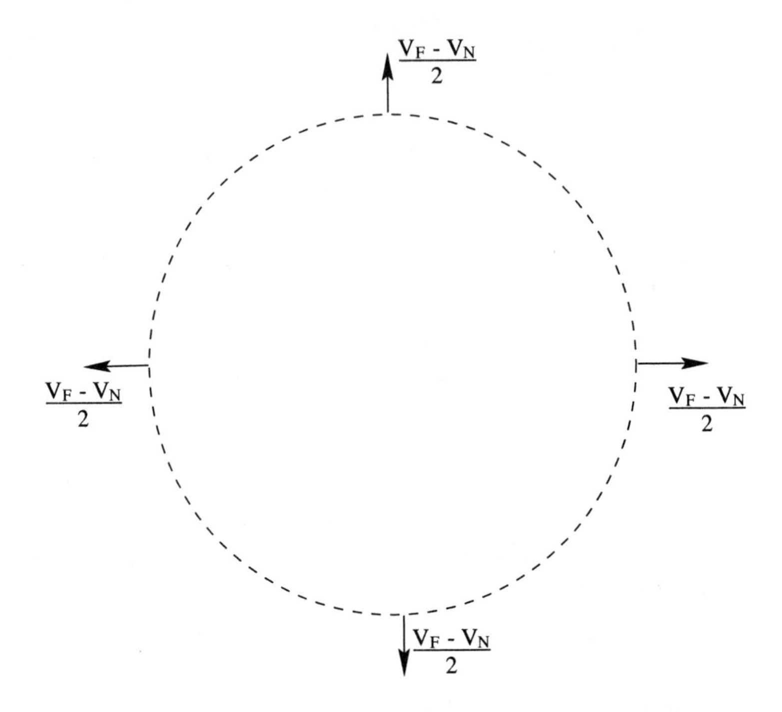

Figure 9.3: Presumed velocity dispersion based on the situation shown in Fig.9.2

moving frame[3] will be much less. As a result the virial mass (necessary to keep the system gravitationally bound) will be much less and the concept of a very large amount of dark matter is unnecessary.

In this section a method has been proposed to extract the true velocity dispersion from the gross redshift data in the case of spherically symmetric rich galaxy clusters. We assume that the proposed theory of velocity-dependent inertial induction, which yielded consistent results in a number of cases discussed in the previous chapters, is correct. The Coma and Perseus clusters will be analysed in greater detail.

9.2.1 Determination of true velocity dispersion

The method for separating the true velocity dispersion from the gross redshift data assumes the rich clusters to be spherically symmetrical, as mentioned above, and that the universe is quasistatic (*i.e.*, there is no overall expansion). The photons are assumed to be subjected to inertial drag according to the model of velocity dependent inertial induction. It has already been shown that a photon suffers a loss of energy as it trav-

[3]It should be further noted that the velocity of the co-moving frame attached to the C.G. of the cluster is also not given by (9.1). In a quasistatic model of the universe, such systems have small random motions.

els through space. Equation (6.10) yields an expression for the resulting redshift, as follows:

$$z = \exp(\frac{\kappa}{c}x) - 1$$

When x is not very large and $z << 1$, the above relation can be simplified yielding

$$z \approx \frac{\kappa}{c}x. \tag{9.3}$$

It was also mentioned in Chapter 6 that since the redshift is generated by an interactive mechanism, it is reasonable to assume that the intensity of the mechanism (and, therefore, the value of κ) is dependent on the matter density in the path of a photon. However, to begin with, let us ignore the local variation of κ and assume it to be equal to

$$\sqrt{\pi \rho G_0}$$

as derived in Chapter 6.

The redshift of a galaxy in a cluster can be said to have two components (the magnitudes of the intrinsic redshifts are comparatively small and, therefore, neglected) as follows:

1. Cosmological redshift proportional to the distance of the galaxy, z_c.

2. Doppler red (or, blue) shift due to the line-of-sight component of the proper velocity of the galaxy, z_v.

Thus, z_c can be taken as a distance indicator (in a quasistatic universe) and when z_c (to some scale representing the distance) is plotted against the distance from the cluster centre with the same scale) the diagram will be of approximately semicircular shape, as shown in Fig. 9.4. This is because it will be the same as the diagram obtained by folding the circle (representing the projection of the spherical cluster on a plane containing the line-of-sight) about the diameter along the line-of-sight. The total redshift of a galaxy is

$$z = z_c + z_v. \tag{9.4}$$

As z_v is due to the component of the velocity along the line-of-sight, when z is plotted against the distance from the cluster centre the diagram will be elongated along the line-of-sight. This is because many galaxies near the boundary of the far side of the cluster have velocities away from the observer, causing the upper quarter circle (of the semi-circle shown in Fig.9.4 to move up by an amount corresponding to $z_v(= v/c$, where v is the velocity of these galaxies). Similarly, many galaxies near the cluster boundary at the near end move with velocity v towards the observer. Hence, a corresponding blueshift of magnitude z_v will cause the bottom quarter of the semi

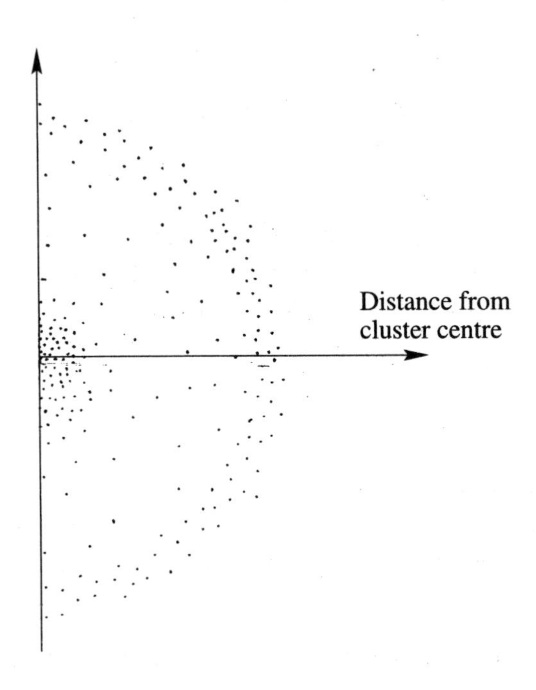

Figure 9.4: Distribution of redshifts of the galaxies in the Coma cluster.

circle in Fig.9.4 to move down by an amount z_v. The velocities of the galaxies near the boundaries in two sides will have very small components along the line-of-sight. Thus, no red or blueshifts will result, and no distortion in a direction normal to the line-of-sight is produced. The plot of the total redshift (in a suitable distance scale) against the distance from the cluster centre will have the appearance of two split quarter-circles as shown in Fig.9.5. The order of magnitude of the true velocity v is given by cz_v, where z_v is determined from the amount of the split of the quarter circles, as indicated in Fig.9.5.

9.2.2 Effect of local variation in κ and shape distortion

The matter density at the core of a cluster is much higher than the average matter density of the universe. Thus, the drag on a photon passing through the core region of a cluster is much greater than that on a photon moving through the inter-cluster space. As a result, a trajectory through a distance x in the core region will result in a redshift which is higher than that when the photon moves through the same distance d outside the core region. Figure 9.6 shows two diametrically opposite galaxies A and B where

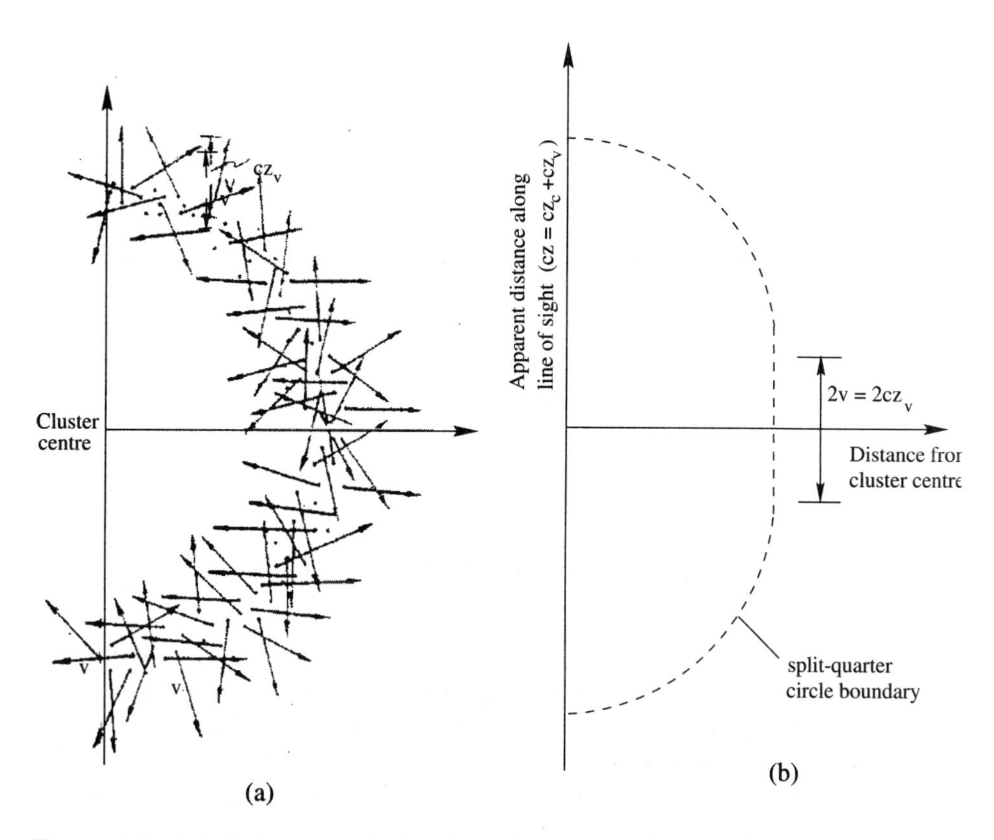

Figure 9.5: (a) Motion of galaxies in a rich cluster (only galaxies near the boundary are shown. (b) Apparent split-quarter-circle shape of the cluster.

the line-of-sight intersects the outer boundary. The distance between A and B is equal to D (which is the diameter of the cluster). If the redshifts of photons from A and B due to inertial drag effect are z_{c_A} and z_{c_B}, respectively, then

$$z_{c_B} - z_{c_A} = \kappa_{BA}.D, \tag{9.5}$$

where κ_{BA} is the local average value of κ when a photon travels from B to A. Next, let us consider two galaxies P and Q where the line of sight intersects the boundary of the core region, as shown. The distance between P and Q is d (= the diameter of the core). We get a relation similar to (9.5) as follows

$$z_{c_Q} - z_{c_P} = \kappa_{QP}.d, \tag{9.6}$$

where z_{c_P} and z_{c_Q} are the redshifts of photons from P and Q due to inertial drag and κ_{QP} is the local average value of κ for a photon going from Q to P. Considering the

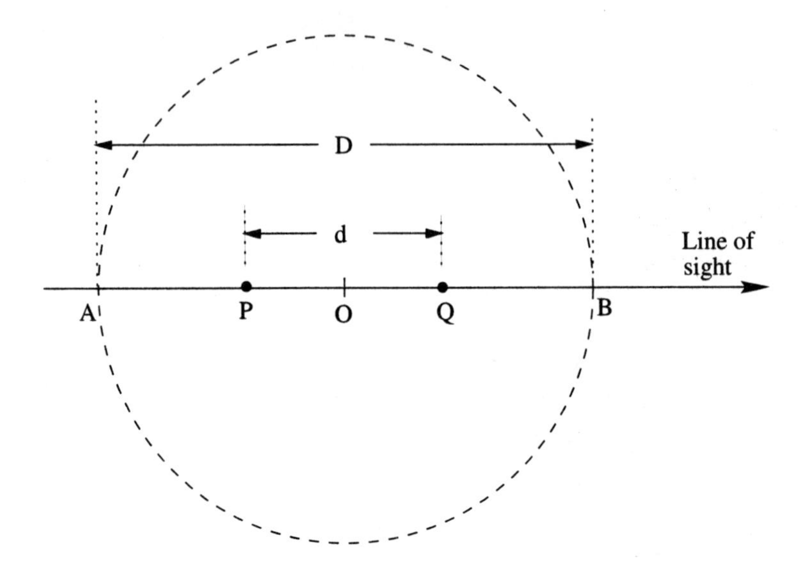

Figure 9.6: Explanation of shape distortion.

data for a number of clusters, it is found that $\kappa_{AB} \approx \kappa$, though κ_{QP} is substantially higher than κ. As a result, the spherically symmetric core, when plotted using z_c as the distance indicator, appears to be elongated along the line of sight. But since $\kappa_{AB} \approx \kappa$ the outer boundary of the cluster remains undistorted once the redshift due to Doppler effect is subtracted from the gross redshift data. The above statements can also be shown in mathematical form as follows:

$$d_{ap} = \frac{z_{cQ} - z_{cP}}{\kappa}, \tag{9.7}$$

where d_{ap} is the apparent diameter of the core determined from the redshift data. But the actual diameter d is given by the following relation:

$$d = \frac{z_{cQ} - z_{cP}}{\kappa_{QP}}, \tag{9.8}$$

since $\kappa_{QP} > \kappa$, $d_{ap} > d$.

On the other hand, when we consider galaxies A and B on the outer boundary of the cluster, we can write the following relations:

$$D_{ap} = \frac{z_{cB} - z_{cA}}{\kappa} \tag{9.9}$$

and

$$D = \frac{z_{cB} - z_{cA}}{\kappa_{BA}}, \tag{9.10}$$

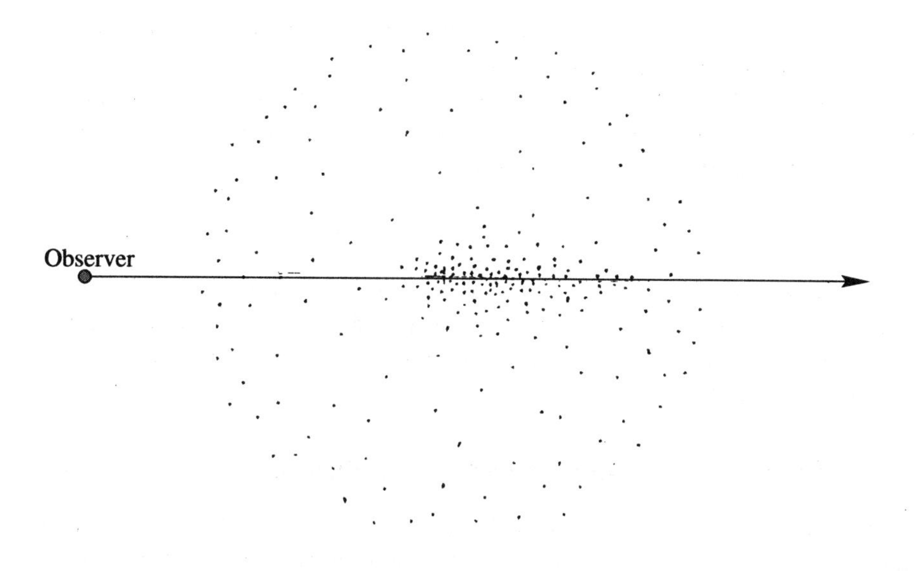

Figure 9.7: Observed cluster shape.

where D_{ap} is the apparent diameter of the cluster determined from the redshift data. Since $\kappa_{BA} \approx \kappa$, it is obvious from the above equations that $D_{ap} \approx D$, and almost no distortion takes place (except that due to the redshift produced by the Doppler effect, which, in any case, is removed in plotting the diagram). Figure 9.7 shows the apparent shape of the cluster derived from the redshift data. It should be noted that the mean position of the core also gets shifted away from the observer, as the elongation takes place only on one side of the core, *i.e.*, the side furthest from the observer.

In recent years cluster masses have also been estimated by using the gravitational lensing phenomenon. Some researchers[4] claim that virial masses are in good agreement with the lensing masses. However, more extensive studies are required before the claim can be substantiated with confidence.

9.2.3 Dependence of apparent magnitude on redshift

Since a predominant fraction of the observed redshift is proposed to be due to velocity-dependent inertial induction, a larger redshift, in general, also implies a larger distance. Assuming the average intrinsic luminosity of the galaxies of a particular type to be constant, the apparent magnitude is expected to increase with redshift. However, it is, unfortunately, difficult to observe this effect distinctly due to the smallness of the effect and the large scatter in the intrinsic luminosities of the galaxies of a particular

[4]Wu, X.P. and Fang, Li-Zhi, *astro-ph*/9701196, 1997.

type in a cluster. In the case of spiral galaxies, the scatter is minimum, and so this type of galaxies is most suitable for observing the effect predicted above.

If two spiral galaxies A and B, with equal intrinsic brightness, are at distances x_A and x_B, $(x_A > x_B)$ from the observer, respectively, their apparent magnitudes, m_A and m_B, should satisfy the following relation[5]

$$m_A - m_B = 5\log_{10}(x_A/x_B) \tag{9.11}$$

Thus, if we plot m versus $\log_{10}x$ the above relation represents a straight line. Since the ratio x_A/x_B is small in the case of galaxies confined to a particular cluster, we can plot m versus x from (9.11) and compare the result with the plots of m versus x for the galaxies in a cluster.

9.2.4 Analysis of Coma and Perseus clusters

Both the Coma and Perseus cluster are rich clusters of galaxies and have been studied quite extensively.[6] Coma is a very clean system with spherical symmetry. On the other hand, the extent of Perseus is not that well defined. Figure 9.8 shows a plot of the line-of-sight distance against the distance from the core centre assuming the value of the constant κ to be equivalent to 50 km s^{-1} Mpc^{-1}. It should be noted at this point that the final result does not depend on the choice of κ, as both the line-of-sight distance scale and the scale for the distance from the cluster centre depend on this choice. The shape is not altered and κ acts as a scale of the plot. A numerical value of κ has been assumed primarily for convenience of representation. From the plot shown in Fig.9.8, it appears that certain galaxies in the foreground and in the background do not belong to the spherical cluster. So far they have been treated as members of the spherically symmetric Coma cluster because of their small angular distance from the centre of the cluster.

The split-quarter-circle configuration is quite obvious from the plot. Figure 9.9 shows how the plot is used to determine the true velocity dispersion of the galaxies

---

[5]The unit of apparent magnitude, m, is a logarithmic measure of brightness. A reduction of 1 magnitude corresponds to an increase in brightness by a factor of 2.5. So,

$$m_A - m_B = 2.5[\log_{10}(\text{brightness of B}) - \log_{10}(\text{brightness of A})].$$

Now, the brightness reduces as the square of the distance. Hence

$$\log_{10}(\text{brightness of B}) - \log_{10}(\text{brightness of A})$$
$$= 2\log_{10}[((\text{distance of A}) / (\text{distance of B})].$$

So,

$$m_A - m_B = 5\log_{10}(x_A/x_B).$$

[6]Kent, S. M. and Gunn, J. E., *The Astronomical Jr.*, V.87, 1982, p.945; Kent, S. M. and Sargent, W. L. W., *The Astronomical Jr.*, V.88, 1983, p.697.

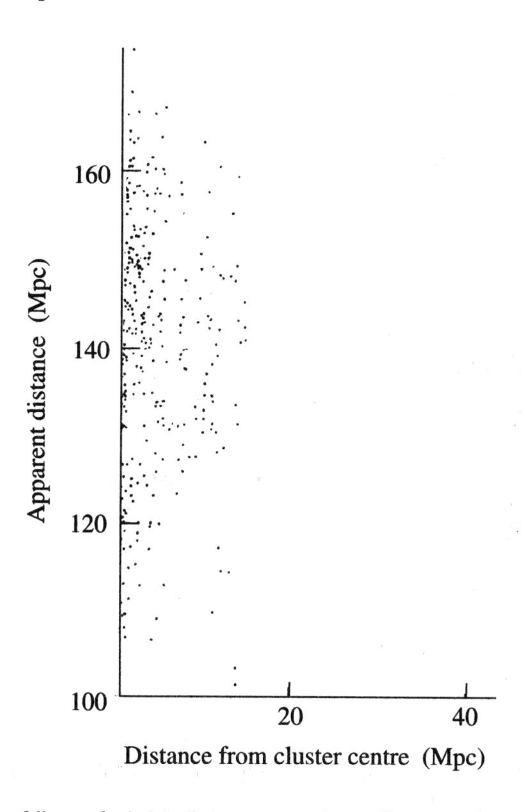

Figure 9.8: Plot of line-of-sight distance against distance from the cluster centre.

in the cluster. The procedure developed in this section yields an estimate of about 350 km s^{-1} for the true velocity dispersion of the system. Comparing this with the conventional value of about 1100 km s^{-1}, we find that the estimate of the virial mass of the system becomes much more realistic. Application of the virial theorem yields a mass/luminosity ratio of about 30 (instead of 180 as estimated in the conventional manner), which is much closer to the mass/luminosity ratio for individual galaxies.

Figure 9.10 shows the m-x plot for the spiral galaxies in Coma which agrees well with the relationship predicted in (9.11). Arp[7] has also obtained a similar dependence of m on z. Figure 9.9 also shows the elongation of the core region, as expected, because κ in the core region is a few times higher than that used to convert the redshift into distance. As mentioned earlier, it is not only the shape which is elongated, the mean position of the core is also shifted away from the geometric centre of the cluster, as predicted.

[7] Arp, H., *Astronomy and Astrophysics*, 1988, V.202, p.70.

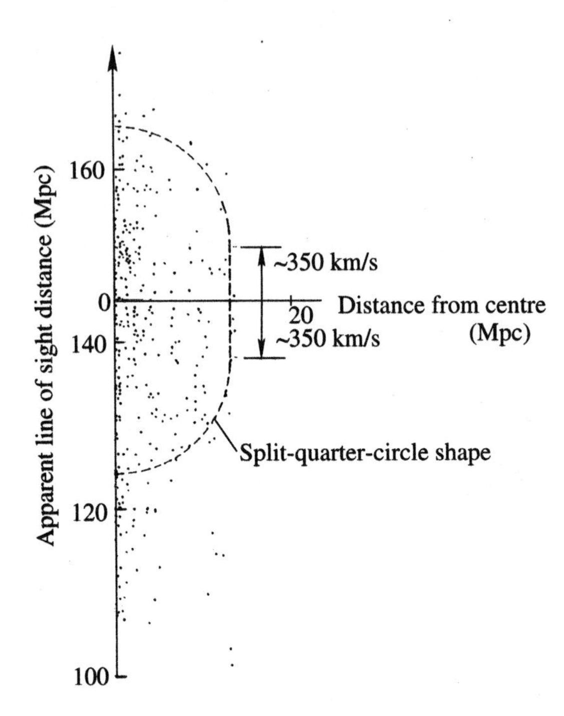

Figure 9.9: Determination of the true velocity dispersion of the Coma cluster.

An analysis of the data from Kent and Sargent[8] indicates the possibility that Perseus could in reality be two smaller clusters, one behind the other along the line-of-sight. With the redshift data converted into apparent distances, this information is plotted against the distance from the core centre (to the same scale). The result is shown in Fig. 9.11. There are two split-quarter-circles representing two spherical clusters. The velocity dispersions of these clusters are on the order of 380 km s^{-1} and 120 km s^{-1} as indicated in the figure. If the conventional high value of 1200 km s^{-1} is replaced by 380 km s^{-1}, the mass-luminosity ratio M/L drops from the conventional value of 300 to about 30, again of the same order as that in individual galaxies. An m-z plot for the galaxies in Perseus is difficult, as adequate data on spiral galaxies of this cluster are not available.

Based on the analysis of the observational data for the two well-studied clusters—Coma and Perseus—using the proposed procedure, an estimate of true velocity dispersions can be made. If the redshifts of the galaxies in clusters are assumed to represent primarily the distance (rather than velocity) because of the drag due to velocity-dependent inertial induction, it is possible to limit the M/L ratio to the same order of

[8]Kent, S. M. and Sargent, W. L. W., *The Astronomical Jr.*, V.88, 1983, p.697.

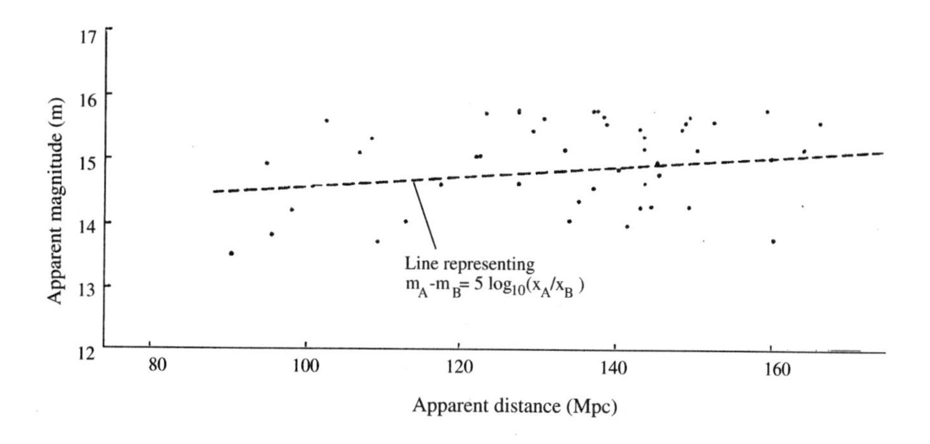

Figure 9.10: Apparent magnitude versus apparent distance.

magnitude as M/L ratios obtained for single galaxies. The observed typical elongation along the line-of-sight is also explained by the mechanism.

9.3 A Concept of Potential Energy in an Infinite Universe

Since velocity-dependent inertial induction leads to velocity-dependent drag on a moving object, the concept of gravitational potential energy of two particles at a distance r from each other appears meaningless. The work done to bring one particle from infinity to a distance r from the other particle depends not only on r but also on the history of the velocity of the moving particle. Therefore, strictly speaking, the system is not conservative. It is needless to mention that a similar situation results because of the acceleration-dependent term also. This poses a major problem for an elegant analysis of the system presented in the previous chapters, and many standard results of gravitationally bound conservative systems are not applicable because of the inertial induction effects. It is also known that the potential energy at a point in an infinite homogeneous universe is $-\infty$ according to conventional physics, which as already mentioned in Chapter 2, is a paradox.

If we consider the universe to be quasistatic, with the velocity and acceleration of all objects small, it may be possible to derive a concept of potential energy.[9] When a

[9]Ghosh, A., *Dynamical Inertial Induction and the Potential Energy Problem*, Apeiron, V.2, n.2, 1995, p.38.

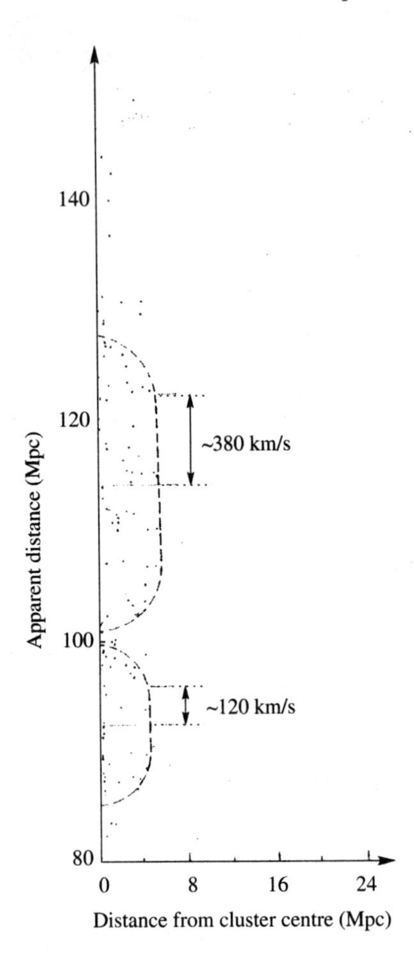

Figure 9.11: Determination of the true velocity dispersion of the Perseus cluster.

particle is brought from infinity to a point at a distance r from another particle, with the velocity and acceleration being infinitesimally small,[10] it is possible to ignore the inertial induction effects throughout, and estimate the work done, as shown below.

The gravitational potential energy of a system of two particles of masses m_1 and m_2 with a separation of r can be defined as the negative of the work done in taking one of the particles away from the other to infinity with infinitesimally small velocity and acceleration. This is given by the following expression:

[10]Of course, the time involved will be infinite.

$$
\begin{aligned}
U &= -\int_r^\infty \frac{Gm_1m_2}{x^2}dx \\
&= -\int_r^\infty \frac{G_0\exp(-\kappa x/c)m_1m_2}{x^2}dx \\
&= -\frac{G_0m_1m_2}{r}\left[e^{\kappa/c}r - \frac{\kappa}{c}r.\mathrm{Ei}(\frac{\kappa}{c}r)\right], \qquad (9.12)
\end{aligned}
$$

where

$$
\mathrm{Ei}(\beta) = \int_\beta^\infty \frac{\exp(-\xi)}{\xi}d\xi.
$$

Using the above formulation, the potential energy of a particle with mass m due to the matter of the universe contained in a spherical shell of radius r and thickness dr, with the particle at its centre, is

$$
dU = -\frac{G_0m}{r}\left[e^{\kappa/c}r - \frac{\kappa}{c}r.\mathrm{Ei}(\frac{\kappa}{c}r)\right]4\pi r^2\rho dr, \qquad (9.13)
$$

where ρ is the mass density of the universe. According to the conventional gravitational law (with G as constant), the potential of a particle due to an infinite universe tends to infinity, as already mentioned. But in the proposed model of inertial induction G has been shown to decrease exponentially with distance.

Because of this, the potential of a particle in an infinite, homogeneous universe remains finite, and can be estimated as follows:

$$
U = \int_{r=0}^\infty dU,
$$

where dU is given by (9.13). Thus,

$$
\begin{aligned}
U &= -4\pi G_0\rho m\int_0^\infty \left[r\exp(-\frac{\kappa}{c}r) - \frac{\kappa}{c}r.\mathrm{Ei}(-\frac{\kappa}{c}r)\right]dr \\
&= -\frac{4\pi G_0\rho mc^2}{\kappa^2}\left[1 - \frac{1}{3}\Gamma(3)\right] \\
&= -\frac{4\pi G_0\rho mc^2}{3\kappa^2}. \qquad (9.14)
\end{aligned}
$$

From (5.7) we know that

$$\kappa = (\chi G_0 \rho)^{1/2},$$

and using this expression in (9.14) we get

$$U = -\frac{4\pi}{3\chi} mc^2. \tag{9.15}$$

With $f(\theta) = \cos\theta.|\cos\theta|$ as assumed throughout the work

$$\chi = \pi,$$

and the expression for the potential energy becomes

$$U = -\frac{4}{3} mc^2. \tag{9.16}$$

Though the physical implication of this energy expression cannot be explained here and now, it is very interesting to note that it is quite near to (in magnitude) mc^2 which represents the energy contained in the particle according to Einstein's famous equation $E = mc^2$.[11] It is felt that the total energy content of the universe is nil and the balance $-1/3mc^2$ is neutralised by the energy of the radiation present in the universe. In fact, such a situation in which the total energy of the universe becomes nil has already been suspected by others.[12] Another way in which the balance $-1/3mc^2$ can accounted for is to attribute it to the lumpiness of the universe. However, further work is necessary to explain the residual amount.

9.4 The Problem of the Great Attractor

There are a few other phenomena which have still eluded convincing solution, and a final decision is possible only when the true nature of the redshifts of distant objects is unambiguously established. The idea of the Doppler origin of the redshift leads us to many peculiar situations. One very interesting example is the Great Attractor.[13] Recent observational results on the peculiar motions of galaxies have indicated that elliptical galaxies in the direction of the Hydra-Centaurus super-cluster in the southern sky are moving with a coherent velocity towards a definite direction. These motions are thought to have arisen from the pull of a "Great Attractor" (GA). This is supposed to be a huge concentration of mass ($\sim 5 \times 10^{16} M_\odot$). It has been claimed that the

[11] It is still more interesting to note that if we take the inclination effect $f(\theta) = \cos\theta$, $\chi = 4/3\pi$ and the potential energy of a particle with mass m is exactly equal to $-mc^2$, which implies that the total energy content of the universe is nil.

[12] Dressler, A., *et al.*,*Astronomical Jr.*, (Letter), V.313, 1987, p.L37.

[13] Dressler, A. and Faber, S. M., *The Astrophysical Journal* (Letter), V.354, 1990, p.L45.

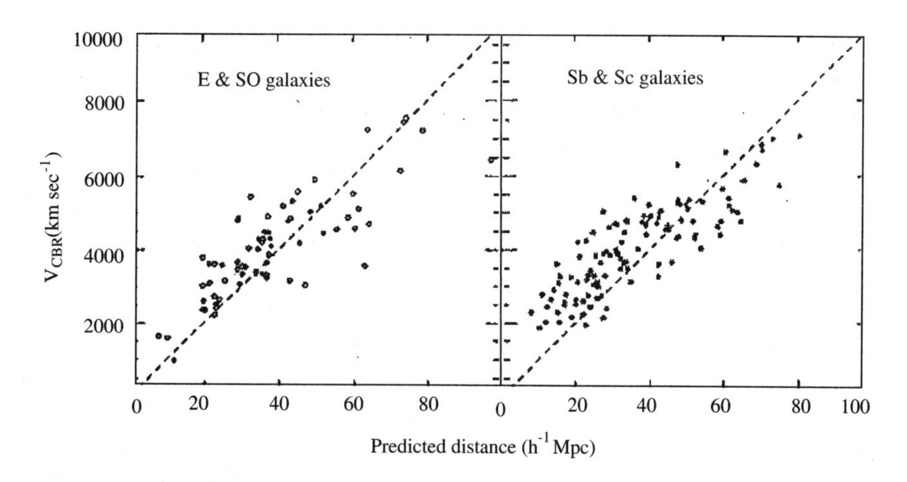

Figure 9.12: (a) The Hubble diagram for 156 E and SO galaxies in the GA region. (b) The Hubble diagram for the new data for 117 spirals in the GA region.

backside in-fall into the GA has also been observed. Figure 9.12 shows the nature of velocity distance relation for the galaxies in the foreground and in the background of the GA. It can be seen from the figure that up to the distance $45g^{-1}$ Mpc, the velocities are larger than the corresponding cosmological velocities. As 45 h^{-1} Mpc is expected to be the location of the suspected GA, the galaxies beyond this location are expected to have velocities lower than the corresponding cosmological velocities due to the pull towards the GA. However, in subsequent work,[14] it has been claimed that there is no evidence of backside in-fall into the Great Attractor. The fresh observational result shows no backside in-fall, and the existence of the GA has been questioned.

9.5 The Nature of the Universe

The theory presented in this monograph has shown that the cosmological redshift is not produced by any expansion of the universe started in a Big Bang, but it is generated by the cosmic gravitational drag. Even in the other commonly discussed model— the theory of a Steady State Universe — the cosmological redshift is assumed to be due to universal expansion of an infinite universe. However, unlike the standard Big Bang cosmology, it proposes no universal evolution nor any temporal or spatial limits to the universe. It is possible to satisfy the Perfect Cosmological Principle through continuous creation of matter maintaining a constant matter density.

The idea of associating velocity with redshift became necessary only because of

[14]Mathewson, D. S., Ford, V. L., and Buchhora, M., *Astrophysical Jr.* (Letter) V.389, 1992, p.L5.

the absence of a mechanism to produce redshift other than the Doppler effect. The gravitational redshift cannot explain the cosmological redshift. A number of mechanisms for producing redshift has been proposed. All these mechanisms are grouped under the common heading "tired light" mechanism. The cosmic drag generated by velocity-dependent inertial induction can also be placed in this group. According to all these theories, photons lose energy (and become redshifted) while travelling through the universe. The drop in energy (and, therefore, the fractional redshift) is proportional to the distance covered, unless the distance involved is extremely large. The major disadvantages with these proposed "tired light" mechanisms (except the one based on velocity-dependent inertial induction) is the fact that they cannot be tested and verified through terrestrial experiments or astronomical observations. On the other hand, the proposed theory has been very impressively vindicated through a number of other independent phenomena. In the preceding chapters, by the application of the model of velocity-dependent inertial induction, a number of long unexplained phenomena have been satisfactorily accounted for.

Many observational programmes have been designed to show direct evidence of universal expansion. All such experiments have failed to arrive at a definite conclusion. There are two observations which are held out as clear proofs of the Big Bang origin of the universe.

In the first case, the relative abundance of certain light elements has been claimed to be correctly predicted by the Big Bang theory. In reality, a number of parameters in the theoretical model are adjusted by using the observational data for certain elements. Then the model is used to predict the abundance of other elements. Though initially an acceptable agreement was achieved, more recent data have created serious problems for fitting the model with them.

The observed 2.7K cosmic microwave background radiation is claimed to be another strong pillar of the standard Big Bang theory. In popular texts it is stated that Gamow and his associates Alpher and Hermann had predicted the existence of a microwave background radiation from their Big Bang model of the universe; and that this predicted background radiation was detected in 1965 by Penzias and Wilson. Thus the Big Bang model was vindicated. However, the existence of a cosmic background temperature has been predicted from a non-expanding quasistatic model of the universe by a number of researchers. Eddington predicted the temperature of the interstellar space to be 3.2K in 1926. He considered the radiation to be in equilibrium.[15] In 1933 Regener[16] derived the value of the background temperature as 2.8K through an analysis of the energy of cosmic rays arriving on the Earth. Nernst followed this work and in 1937[17] proposed a model of an infinite universe without expansion. Considering the

[15]Eddington, A.S.– *The Internal Constitution of the Stars*, Cambridge University Press, 1926. (Later Eddington changed his view and accepted an expanding model of the universe).

[16]Regener, E., *Zeitschrift für Physik*, V.80, p.666, 1933.

[17]Nernst, W., *Zeitschrift für Physik*, V.106, p.633, 1937.

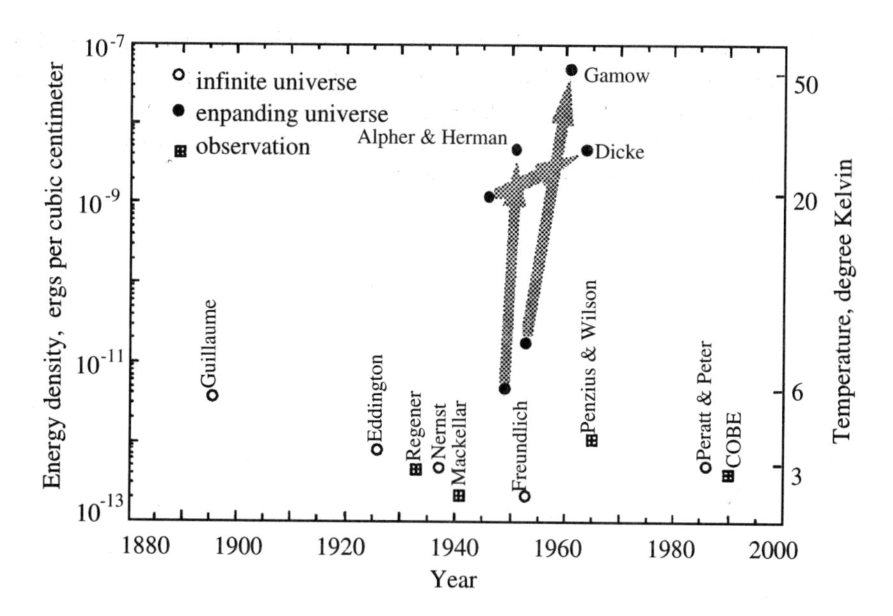

Figure 9.13: History of the blackbody cosmic background radiation over 100 years.

background temperature to be 2.8K he tried to derive the light absorption by the cosmic medium, which resulted in the cosmological redshift. In 1954 Finlay-Freundlich also tried to explain the cosmological redshift by an interaction in the interstellar medium. In his work, Freundlich derived a blackbody temperature of the intergalactic space which lies between 1.9K and 6K. This was supported by Max Born, and he concluded: "Thus the redshift is linked to radio-astronomy." On the other hand, in 1949 Alpher and Hermann derived the value of the cosmic background radiation temperature to be more than 5K from the Big Bang model. This was modified to 7K by Gamow in 1953. In 1961 Gamow further modified this temperature to 50K. Figure 9.13[18] shows diagrammatically the true history of the black body cosmic background radiation. In light of this history, the strongest "proof" of the Big Bang model turns out to be its strongest counter-evidence.

The primary objective of this section is to remind the reader that the issue of the cause of the cosmological redshift is still unresolved. Although mainstream cosmologists generally accept the expansion hypothesis, to date there is no unequivocal evidence in support of this theory.

The theory of velocity-dependent inertial induction quite successfully explains the cosmological redshift. The theory is further vindicated by a number of other phenom-

[18]Peratt A., In: *Plasma Astrophysics and Cosmology*, Kluwer Academic Publishers, 1995.

ena, unlike the other proposed models of "tired light" mechanisms. The large number of agreements between the quantitative deductions of the proposed theory with firmly established observational data should, therefore, encourage other researchers to undertake further investigations into other cases. Attempts should be made also to detect the predicted secular retardation of the rotation of Mars.

Epilogue

When I find high school students nowadays solving mechanics problems involving pulleys, inclined planes, rockets, cars, I cannot but help think of the early summer of 1956. I had just completed my high school in a remote village of Bengal and was waiting for my admission to the district college for the Intermediate Science programme. My father thought that the time might be better utilized if I were to get some prior exposure to science. In those days, up to Class-10 there were hardly any science topics in the programme, and we had absolutely no introduction to mechanics. Even the terms like velocity, acceleration, momentum, etc. were totally unfamiliar to us in the high school. One of my cousins had finished his Intermediate Science and was a trainee in a steel plant. He came to spend a few weeks at our home, and first introduced me to the names of Newton and Galileo. He gave me my first ever lesson in elementary kinematics, the parallelogram laws of the addition of forces and motion parameters. Soon afterwards, I was introduced to the laws of motion by another young postgraduate in Mathematics from the village. By then he had left Mathematics and was studying Law, but had returned to spend his summer vacation at home.

I remember the tremendous mental block I had in conceiving of the basic concepts. By that time, I was familiar with multiplying physical quantities by numbers. Somehow, ideas of velocity and acceleration, which involved length and time, I could grasp. What was very difficult for me at that time was to conceive of the idea of one physical quantity being multiplied by another physical quantity. For me the stumbling block was the concept of momentumthe product of mass and velocity. I can still remember the utter exasperation of the young law student who had already completed a Master of Science in Mathematics from Calcutta University. He was completely baffled by my difficulty. It took a long time for me to accept the concept of momentum.

My next hurdle was to swallow the idea that objects resist acceleration but not velocity. This was a larger problem, which, quite frankly, I could not overcome, even in later years. In particular, I could not understand why acceleration has special status, and why the velocity of an object does not need any support to remain constant. Much later, when I read the history of mechanics, I realized that my doubts were not entirely unjustified, as it was earlier thought even by scientists and philosophers that an impetus was needed for an object to move with a constant velocity.

There is no obvious reason why an object moving through empty space does not experience any resistance. The only reason why this is thought to be the case may be the fact that the resistance is extremely small compared to the resistance encountered in the case of acceleration. If we assume a situation where all objects are charged, then we would never have been able to detect gravitation. Gravitation is so weak a force in comparison with electromagnetic and electrostatic forces that its effects are completely overwhelmed. Cannot a similar situation be true in the case of a bodys resistance to both acceleration and velocity? The reader of this monograph has by now seen that it is possible for a velocity-related force to exist, although it is extremely small. But as mentioned earlier, this realization has one very profound consequence. The cosmological redshift is caused by this extremely small force, and this fact drastically alters our model of the cosmos.

The matter of an exact equivalence between the inertial and gravitational masses of an object is another intriguing manifestation of nature. Machs Principle alone does not explain this equivalence. The exceptional precision of the equivalence requires extreme fine-tuning of a number of independent parameters in the universe. One possible way to achieve this degree of exactness is for a servomechanism or feedback mechanism to be active. We know from the time of Einstein that gravity is unique in the way that it acts on itself i.e., gravity acts on gravity. This servomechanism, along with the attenuation of gravity due to the presence of matter in the universe (arising from the proposed extension of Machs Principle) does result in an exact equivalence, as shown in the monograph. This removes the gravitational paradox also, as demonstrated and first suggested by Laplace, and later by Seeliger.

Thus the proposed theory resolves two major issues having profound philosophical implications. Acceleration no longer enjoys a qualitatively distinct status. Its importance in our daily experience and in most science and engineering problems may be coincidental, as forces due to other derivatives of displacement are very small in comparison.

I have tried to test the validity of the hypothesis by applying it to a number of other unrelated phenomena. It is a matter of great satisfaction that in each and every case where a detectable effect is expected, it has been found to be present. In many cases attempts had been made to explain these effects with the help of conventional physics; but in all cases the attempts came up against difficulties. In some cases no reasonable explanation had existed at all. I am sure that with rapid advances in technology, our capacity to measure and experiment will reach higher levels of accuracy; it is thus only a matter of time until many new tests will become available.

I am quite aware of the fact that acceptance of a new theory takes time. Therefore, in spite of a large volume of evidence in support of the proposed theory, I am reconciled to the prospect that it may not find quick acceptance. This is especially true given that I have not been able to express my theories through the current formalisms of physics and couch them in the arcane mathematics so much in vogue. Nevertheless, I dare to hope that some readers will agree with me as to the need for a reexamination of basic issues. May the reader find in this monograph an adequate conceptual foundation for this task.

References

1. Arp, H. : *Astronomy and Astrophysics*, Vol. 202, 1988, p.70.

2. Assis, A. K. T. : *Foundations of Physics Letters*, Vol. 2, 1989, p. 301.

3. Barbour, J. B. : *Absolute or Relative Motion*, Cambridge University Press, 1989.

4. Beckeres, J. M. and Nelson, G. D. : *Solar Physics* , Vol. 58, 1978, p. 243.

5. Bekenstein, J. and Milgrom, M: *The Astrophysical Journal*, Vol. 286, 1984, p. 7.

6. Berkeley, G. : *The Principles of Human Knowledge*, Vol. 35 of Great Books of the Western World, Encyclopaedia Britanica, Chicago, 1952.

7. Bertolli, B. et al. in *Gravitation: An Introduction to Current Research*, (ed.) Louis Witten, John Wiley, 1962.

8. Bootlinger, C. F. : *Astronomische Nachrichten*, Vol. 1991, 1912, p. 147.

9. Bray, R. J. and Loughhead, R. E. : *The Solar Granulation,*, Chapman and Hall, 1867.

10. Brown, E. W. : *Monthly Notices of Royal Astronomical Society*, Vol. 63, 1903, p. 396.

11. Calame, O. and Mulholland, J. D., in *Tidal Friction and Earth's Rotation*, (eds.) P. Brosche and J. Sundermann, Springer-Verlag, Berlin, 1978.

12. Cloutman, L. D. : *Space Science Review* Vol. 2, 1980, p. 23.

13. Cohen, R. S. and Seeger, R. J. (eds.) : *Earnst Mach : Physicist and Philosopher*, D. Reidel Pub. Co., Dordrecht, Holland.

14. Cook, A. H. : *Interiors of Planets*, Cambridge University Press, 1980, p. 195.

15. Cure, J. : *Galilean Electrodynamics*, Vol. 2, 1991, p. 43.

16. Dicke, R. H. : "The Many Faces of Mach," in *Gravitation and Relativity*, (ed.) Chin, H. Y. and Hoffman W. F.; W. A. Benjamin, New York, 1964.

17. Dicke, R. H. : in *The Earth-Moon System*, (eds.) B. G. Marsden and A. G. W. Cameron, Plenum, New York, 1966.

18. Dragoni, G. : *Proceedings of the X course on Gravitational Measurements Fundamental Metrology and Constants*, Dordrecht, Kluwer, 1988.

19. Dressler A. et al. : *Astronomical Journal (letters)*, Vol. 313, p. L. 37, 1987.

20. Dressler A and S. M. Faber : *The Astrophysical Journal*, Vol. 354, p. L45-L48, May 10, 1990.

21. Eddington, A. S. : *The Internal Constitution of Stars*, Cambridge University Press, 1926.

22. Felten, J. E. : *The Astrophysical Journal*, Vol. 286, 1984, p. 3.

23. Freeman, K. C. : *The Astrophysical Journal* Vol. 160, 1970, p. 811.

24. Fridman, A. M. and Polyachenko, V. L. : *Physics of Gravitating Systems, I: Equilibrium and Stability,* Springer-Verlag, New York, 1984, p. 327.

25. Fuji, Y. : *General Relativity and Gravitation*, Vol. 6, 1975, p. 29.

26. Ghosh, A. : *Pramana (Journal of Physics)*, Vol. 23, 1984, p. L671.

27. Ghosh, A. : *Pramana (Journal of Phsyics)*, Vol. 26, 1986, p. 1.

28. Ghosh, A. : *Earth, Moon and Planets*, Vol. 42, 1988, p. 69.

29. Ghosh, A. : *Astrophysics and Space Science* Vol. 227, 1995, p. 41.

30. Ghosh, A. : *Apeiron,* Vol. 2, no. 2, 1995, p. 38.

31. Ghosh, A. , Rai, S. and Gupta A. : *Astrophysics and Space Science,* Vol. 141, 1988, p. 1.

32. Grabowski, B. *et al.* : *The Astrophysical Journal*, Vol. 313, 1987, p. 75.

33. Greenstein, J. L. and Trimble, V. L. : *The Astrophysical Journal*, Vol. 149, 1967, p. 283.

34. Gyan Mohan : *Frames of References*, in Lectures delivered on the course of the Tercentenary occasion of Newton's Principia, IIT Kanpur, India, Feb. 27-28, 1987 (Unpublished manuscript).

35. Hagiharia, Y. : *Celestial Mechanics,* Part I, MIT Press, 1972.

36. Hoyle, F. : *Quarterly Journal of Royal Astronomical Society*, Vol. 1, 1960, p. 28.

37. Jaakkola, T. *et al.* : *Nature,* Vol. 256, 1975, p. 24.

38. Jaakkola, T., Karoji, H. et al. : *Monthly Notes of the Royal Astronomical Society*, Vol. 177, 1976, p. 191.

39. Kent, S. M and Gunn, J. E. : *The Astronomical Journal*, Vol. 87, 1982, p. 945.

40. Kent, S. M. and Sargent W. L. W. : *The Astronomical Journal*, Vol. 88, 1983, p. 697.

41. Kropotkin, P. N. : *Soviet Physics Doklade*, Vol. 33(2), 1988, p. 85.

42. Kropotkin, P. N. : *Soviet Physics Doklade*, Vol. 34(4), 1989, p. 277.

43. Kropotkin, P. N. : *Apeiron*, Nos. 9-10, 1991, p. 91.

44. Kuhn J. R. and Kruglyak, L. : *The Astrophysical Journal*, Vol. 313, 1987, p. 1.

45. Küveler, G. : *Solar Physics*, Vol. 88, 1983, p. 13.

46. Laplace, P. S. : "Traite de Mecanique Céleste" in *Oeuvres de Laplace*, Vol. 5, Book 16, Chapter 4, Gauthier-Villars, Paris, 1880.

47. Mach, E. : *The Science of Mechanics — A Critical and Historical Account of the Development*, Open Court, La Salle, 1960 (Orginally published in 1886).

48. Maneff, C. : *Zeitschrift für Physik*, Vol. 34, 1930, p. 766.

49. Majorana, Q. : *Philosophical Magazine*, Vol. 39, 1920, p. 488.

50. Majorana, Q. : *Comptes Rendues de L'Academie des Sciences* (Paris), Vol. 173, 1921, p. 478.

51. Majorana, Q. : *Journal de Physique*, Vol. 1, 1930, p. 314.

52. Mathewson D. S. , Ford, V. L. and Buchhora, M. : *The Astrophysical Journal (Letter)*, Vol. 389, 1992, p. L5.

53. McElhinny, M. W. : *The Earth: Its Origin, Structure and Evolution*, Academic Press, London, 1979.

54. Melchoir, P. : *The Tides of the Planet Earth*, Pergamon Press, London 1978.

55. Merat et al. : *Astronomy and Astrophysics*, Vol. 174, 1974, p. 168.

56. Milgrom, M. : *The Astrophysical Journal*, Vol. 270, 1980.

57. Moffet, T. J. et al. : *The Astronomical Journal*, Vol. 83, 1978, p. 820.

58. Munk, W. H. and Mcdonald, G. L. F. : *The Rotation of the Earth*, Cambridge University Press, 1960.

59. Nernst, W. : *Zeitschrift für Physik*, Vol. 106, 1937, p. 633.

60. Ostriker, J. P. and Deebles, P. J. E. : *The Astrophysical Journal*, Vol. 186, 1973, p. 467.

61. Pechlaner, E. and Sexl, H. : *Communications in Mathematics and Physics*, Vol. 2, 1966, p. 165.

62. Peratt, A. : *Plasma Astrophysics and Cosmology*, Kluwer Academic Publishers, 1995.

63. Pound, R. V. and Rebka, G. A. : *Physical Review Letters*, Vol. 4, 1960, p. 337.

64. Pound, R. V. and Snider, J. L. : *Physical Review*, Vol. B 140, 1965, p. 788.

65. Priest, E. R. : *Solar Magnetohydrodynamics*, D. Reidel Pub. Co., Dordrecht, Holland, 1982.

66. Regener, E. : *Zeitschrift für Physik*, Vol. 106, 1937, p. 633.

67. Rosenberg, G. D. and Runcorn, S. K. : *Growth, Rhythms and the History of the Earth's Rotation*, John Wiley, London, 1975.

68. Rubin, V. C. , Ford Jr. W. K. and Rubin J. S. : *Astrophysical Journal Letters*, Vol. 183, 1973, p. L111.

69. Sadeh et al. : *Science*, Vol. 159, 1968, p. 307.

70. Sanders, R. H. : *Astronomy and Astrophysics*, Vol. 154, 1986, p. 27.

71. Sciama, D. W. : *On the Origin of Inertia*, Monthly notices of the Royal Astronomical Society, Vol. 113, 1953, p. 34.

72. Sciama, D. W. : *The Physical Foundations of General Relativity*, Heinemann Educational Books Ltd. , 1972.

73. Seeliger, H. : *Astronomische Nachrichten*, Vol. 137, 1895, p. 129.

74. Seeliger, H. : *Über die Anwendung der Naturgesetze auf das Universum*, Berichte Bayer, Akad. Wiss., Vol. 9, 1909.

75. Shipman, H. L. , and Sass, C. A. : *The Astrophysical Journal*, Vol. 235, 1980, p. 177.

76. Sinclair, A. T. : *Astronomy and Astrophysics*, Vol. 220, 1989. p. 321.

77. Stacey, F. D. : *Physics of the Earth*, John Wiley, New York, 1977.

78. Tisserand, M. F. : *Comptes Rendues de l'Academie des Science* (Paris), Vol. 75, 1872, p. 760.

79. Toomre, A. : *The Astrophysical Journal*, Vol. 138, 1913, p. 385.

80. Trimble, V. : "Existence and Nature of Dark Matter in the Universe," *Annual Review of Astronomy and Astrophysics,* Vol. 25, 1987, p. 425.

81. van Flandern, T. : *Dark Matter, Missing Planet and New Comets,* North Atlantic Books, 1993.

82. Vessot, R. F. C. et al. : *Phyical Review Letters*, Vol. 45, 1980, p. 2081.

83. von Weizsäcker, C. F. : *Zeitschrift für Astrophysik*, Vol. 22, 1943, p. 319.

84. von Weizsäcker, C. F. : *Zeitschrift für Astrophysik*, Vol. 24, 1947, p. 181.

85. Weber, W. : Leipzig Abhandl. (1846) : Ann. d. Phys. lxxiii, 1848, English translation in Taylor's Scientific Memoirs, 1852, p. 489.

86. Winkler, K. P. : *Berkeley, Newton and Stars : Studies in History and Philosophy of Science,* Vol. 17, 1886, p. 23.

87. Whittaker, E. : *History of the Theories of Aether and Electricity,* Vol. 1 and 2, Thomas Nelson and Sons, 1953.

Index